SCIENTIFIC
INDIANA

DUANE S. NICKELL

THE
History
PRESS

Published by The History Press
Charleston, SC
www.historypress.com

Copyright © 2021 by Duane S. Nickell
All rights reserved

First published 2021

Manufactured in the United States

ISBN 9781540248107

Library of Congress Control Number: 2021934136

To the several thousand students whom I had the privilege of teaching during my thirty-year career. I hope you learned something. Thanks for the memories!

CONTENTS

ACKNOWLEDGEMENTS

Writing is a solitary yet enjoyable and deeply rewarding activity. Crafting this book helped preserve my sanity during a time of political chaos and pandemic danger. There are several people I would like to thank for their help in this endeavor. First, I want to thank my editor, John Rodrigue, for his enthusiastic embrace of this project. I'm glad I was lucky enough to find an editor who appreciates science. Thanks also to the team at The History Press for their work on the cover and their help in editing and improving the manuscript. Finally, a very special thanks to my wife, Karen Markman, for her careful reading and critiquing of the manuscript.

INTRODUCTION

When you think of Indiana, what things come to mind? Automobile racing, basketball and corn, right? The word *science* probably doesn't occur to you. If you were asked to name some famous Hoosiers, my guess is you would cite athletes, actors, musicians, writers, television personalities and political leaders. Would you name any scientists? Probably not. But Indiana has a rich scientific heritage. As we shall see, just a few years after achieving statehood, Indiana was home to a community of scientists at New Harmony. Few states can boast of such a strong scientific presence so early in its history.

What accounts for this state of affairs? After all, scientists are the real heroes of the modern world. Instead of making stuff up, they figure things out. And when you figure out how the universe works, you can apply that knowledge to solve problems and improve lives. Science can cure diseases, find new sources of energy and solve the climate crisis. Science, when applied, creates new technologies like cellphones and computers. Science isn't perfect, but it's the best tool we've got.

The sad fact is that in today's world, science doesn't always get the respect it deserves. This is especially true in the United States, where a significant segment of the population doesn't trust egg-headed intellectuals and experts. Even some of our political leaders address problems by going with their gut instead of following the data. Not enough people understand basic scientific ideas, and some reject well-established scientific principles. Expert scientific opinion is often ignored and sometimes ridiculed. For example, although there is a strong scientific consensus that climate change is a real problem largely caused by the burning of fossil fuels, many people stubbornly refuse to accept it. Too many people simply believe what they want to believe, reason and rationality be damned!

This book is a small step toward solving this big problem. It is a celebration of science and scientists in Indiana. The book begins with a chapter on the dawn of science in the Hoosier State. The remainder of the book consists of biographical sketches of seventeen scientists who are, in some significant way, connected to the state. Each biography begins with the name of the scientist along with their scientific field, major contribution and connection to Indiana, followed by a quote by the scientist. The biographies are presented in chronological order according to birth year. Each biography traces the life of the scientist and includes a simple explanation of their scientific contributions.

What were the selection criteria for the scientists included in this book? The first requirement was a strong connection to Indiana. A significant fraction of their lives had to be spent actually living and working in Indiana or they must have earned a degree from a college in the state. What is a significant fraction? Since the time normally required for earning a degree is four years, I arbitrarily set this as the minimum time. Thus, spending a year or two in the state was not sufficient to justify inclusion in this book.

The second requirement was that the scientist's accomplishments must be of the highest order. One obvious indicator of the significance of scientific work is winning the Nobel Prize. Twelve of the scientists chosen for inclusion in this book are Nobel laureates. But the Nobel Prize is not a perfect indicator of scientific achievement. The Nobel Prize is awarded for discoveries or for theoretical work that has been verified by experiment or observation. It is not awarded for unsubstantiated theories. Also, the Nobel Prize is awarded only in the scientific fields of physics, chemistry and physiology or medicine. There is no Nobel Prize explicitly for biology, although physiology and medicine are encompassed by biology. There is no Nobel Prize for astronomy, although the field of physics was eventually interpreted more broadly to include astrophysics. Thus, five of the seventeen scientists portrayed in this book have not won a Nobel Prize but nevertheless made substantial contributions to science.

Those are the criteria I used to choose the scientists. To be clear, I'm defining science narrowly to include only the natural sciences of biology, chemistry, physics, geology, astronomy and their associated disciplines.

The Nobel Prize Medal. *Shutterstock.*

Natural science does not include engineering, invention or the social sciences. The list is, I freely admit, imperfect and incomplete. It is simply my subjective opinion of who should be included. I regret that all the scientists portrayed in this book are male. I wish I could have justified including Elinor Ostrom, a professor from Indiana University who won the Nobel Prize for Economics and the first woman ever to win the award in that discipline. But alas, economics is not a natural science. The sad fact is that until recently, women were not afforded the same educational or professional opportunities as men. That situation is slowly changing as more and more women enter scientific fields.

I hope you enjoy reading this book as much as I enjoyed writing it. Let us begin!

I

SCIENCE IN INDIANA BEGINS IN NEW HARMONY

Science in Indiana began in a place called New Harmony, in the southwestern part of the state on the banks of the Wabash River. It was there in 1814 that a splinter group of German Lutherans formed a religious commune called Harmonie. The sect had come to America from Germany in 1804, settled in Pennsylvania for a few years and then moved on to Indiana territory. Their authoritarian leader was George Rapp, a fundamentalist preacher who believed that Jesus Christ would soon return and life as we know it would end. Procreation was thus rendered unnecessary, so Rapp insisted that all members of his flock, married or single, remain celibate. The Rappites, as they were called, perhaps refocusing their pent-up sexual energy, spent the next ten years building a successful and prosperous community. They cleared two thousand acres of land and built forty houses, two churches, a school, a library, a sawmill, a distillery, a greenhouse, a gristmill and a five-story stone granary. They grew crops, planted an orchard and cultivated a vineyard. The commune was economically successful, selling a variety of goods including beer, wine, whiskey and woolens.

After a decade of exhausting work, Rapp suddenly decided to sell Harmonie in 1824 and move back to Pennsylvania, where he and his followers would establish a new commune near Pittsburgh. Why did Rapp decide to move? The reason he gave was that the German settlers did not like the weather. But the weather in western Pennsylvania is not much different from Indiana. He also complained that the commune was too far from the centers of business. Perhaps. But some have suggested that the commune was becoming a little too prosperous and, as a consequence, harder for Rapp to control. For whatever reasons, Rapp put a "for sale" sign up on his Indiana community.

Robert Owen, a wealthy Welsh textile manufacturer, philanthropist and social reformer, became interested in the property. Owen had been a part-owner and manager of a large textile mill in Scotland, where he greatly improved the working conditions. In America, he saw an opportunity to build a new experimental community based on the ideas of socialism. He was an outspoken critic of established religion, the institution of marriage and the ownership of private property. These views were anathema to most Americans, but he was such a charismatic and persuasive speaker that most listeners overlooked his radical opinions. In October 1824, Owen and his son William sailed to America to inspect the community of Harmonie.

Scottish philosopher Robert Owen, who established a utopian community at New Harmony, Indiana. *Shutterstock*.

Owen arrived in mid-December and spent a week looking over the property. On January 3, 1825, he bought the entire town for the sum of $150,000. The Rappites packed up and left for Pennsylvania; Owen renamed the town New Harmony and vowed to turn the former religious settlement into a secular utopia. But rather than attending to the details of organizing the new settlement, Owen immediately left on a three-month speaking tour to spread the word about his socialist settlement and round up recruits. The tour included a stop in Philadelphia, where in March, he met with members of the Academy of Natural Sciences, several of whom were interested in joining Owen in New Harmony. Owen was eager for the scientists to join his colony and transform it into a major center for science. The Philadelphia scientists were attracted by the idea of a utopian society where science would occupy its rightful place at the pinnacle of human endeavor.

The academy was financially supported by a Scottish geologist, educator and philanthropist named William Maclure, and the scientists would need his permission to go to Indiana. Maclure made his fortune in the mercantile business in Scotland, retiring at age thirty-four to pursue his own interests. While living in Switzerland, he became a disciple of Johann Pestalozzi, a

liberal-minded Swiss education reformer, and vowed to build a system of schools based on Pestalozzi's pedagogical methods. In 1807, he took it upon himself to make a geological survey of the United States and spent years crisscrossing the country and visiting nearly every state. The result was the first geological map of the United States. For this and other contributions, Maclure is known as the "father of American geology."

Eventually, Owen convinced Maclure to join him in New Harmony and bring some of the academy scientists with him. Maclure invested $10,000 in the venture, and the men agreed that Maclure would be in charge of science and education while everything else would be up to Owen. Once the decision was made, things happened fast; within a few weeks, the group was ready to go. The scientists included Gerhard Troost, a Dutch chemist and crystallographer; Charles-Alexandre Lesueur, an artist and naturalist; and Thomas Say, an entomologist and conchologist. These were Indiana's first scientists. The lead teacher was Marie Duclos (Madame) Fretageot, a French-born woman who was an expert in Pestalozzian pedagogy and a close friend of Maclure.

On November 28, 1825, Maclure and his group left Philadelphia, arriving by coach in Pittsburgh on December 8. From there, thirty-five passengers and ten crew took a keelboat down the Ohio River toward Indiana. The boat's official name was the *Philanthropist*, but Owen dubbed it the "Boatload of Knowledge" in honor of the learned passengers. After running aground and being stuck in ice, the boat finally arrived in New Harmony in January 1826. By another route, Maclure sent books and scientific apparatus so that his educational experiment could begin immediately.

This time, Owen remained in New Harmony for eighteen months—his longest stay. At first, everything appeared to be working well and hopes in the community were high. In Maclure's schools, run by the Education Society, the students learned through a combination of intellectual effort and physical labor. They received instruction in language, mathematics and nature study. The boys made shoes, hats, pottery and brooms for the community while the girls washed, cooked and worked in the textile mills. Soon, they had workshops for tailors, carpenters and weavers. Madame Fretageot trained the teachers and taught students ages two to five in what was probably the first preschool in America. The scientists provided instruction in their fields of knowledge. But some aspects of the schools were downright Dickensian. Student dress was simple, the diet was spartan and the days were long, starting at 5:00 a.m. and ending at 8:00 p.m. And, in a policy that by today's standards seems cruel, students were kept separate from their parents. Nevertheless, most students later recalled their school days with fondness.

But soon, there was growing disharmony in New Harmony. Although the schools and the scientists were successful, the town's governance was in disarray and went through several reorganizations and a handful of constitutions. To top it all off, on July 4, 1826, the fiftieth anniversary of the Declaration of Independence, Owen delivered a speech titled "Declaration of Mental Independence" in which he repeated his attacks on religion, marriage and private property. This was nothing new to the New Harmonites, who had heard it all before. But the speech was widely reported and denounced by newspapers across the country, earning New Harmony a reputation as an evil den of atheism and sexual promiscuity. The relationship between Owen and Maclure deteriorated, and their arguing escalated to the point where legal action was taken by both sides. Maclure even attempted to have Owen arrested, a fate that Owen only narrowly avoided.

Less than a year after Owen's infamous speech, the utopia disintegrated into chaos. On May 6, 1827, Robert Owen spoke to the citizens of New Harmony for the last time before heading back to England. He had lost most of his fortune in pursuit of his dream. After Owen left, the socialist experiment in New Harmony ended and capitalism took over; the land was subdivided and put up for sale, the houses were rented and stores opened for business.

Robert Owen's grand experiment had failed, but his legacy lived on through his eight children, three of whom became leaders in the fields of science, education and politics. Two of his sons, David Dale Owen and Richard Dale Owen, became geologists, probably due to the influence of Maclure and the other scientists at New Harmony. Richard Dale would play a role at both of Indiana's flagship universities. He joined the faculty at Indiana University on January 1, 1864, after serving as an officer in the Civil War. He chaired the natural science department and taught classes in geology, chemistry, language and natural philosophy. While maintaining his position at Indiana University, he became the first president of Purdue University in 1872, but disagreements with the trustees ended with his resignation in 1874 before any classes had been held. Buildings on both campuses are named in his honor.

Many of the scientists remained in New Harmony and continued to do research, publish and teach. Thomas Say wrote two classic works, *American Entomology* and the seven-volume *American Conchology*; Charles-Alexandre

View of New Harmony from the visitor's center. *Paul Hein/iStockphoto.*

Lesueur published *Fish of North America*, the first description of fish in the Great Lakes; and François Michaux produced *North American Sylva*, a guide to the trees of the United States and Canada. Earlier, Maclure had bought a printing press and copper plates for engraving so that books could be published in New Harmony at a fraction of the usual cost. The books were then sold by peddlers traveling by wagon.

During the 1840s and 1850s, New Harmony remained an intellectual center and was filled with scientists and other educated workers. After that, the town's influence waned, its great specimen collections were transferred to big-city museums and it no longer attracted scientists and naturalists. New Harmony became a small, isolated and largely forgotten midwestern town. But the failed utopian experiment at New Harmony succeeded in bringing science to Indiana.

Today, most scientific research in Indiana and in the United States is done at large research universities. Indiana has six campuses that are classified as research universities: Indiana University, Purdue University, Ball State University, Indiana University Purdue University at Indianapolis (IUPUI), Indiana State University and the University of Notre Dame. Faculty at smaller colleges also contribute to scientific research. Scientific research isn't free; it costs money to buy equipment and pay assistants. Where does the money come from to fund the research? Various agencies of the federal government, such as the National Science Foundation (NSF) and the National Institutes of Health (NIH), provide most of the funding for scientific research. Of course, the federal government gets its money from taxpayers. About two cents out of every federal tax dollar goes to support scientific and medical research.

That's a brief look at how science began in Indiana. Now let's turn our attention to the people who actually do the science. Come with me as we explore the lives of some of the greatest scientists in Indiana history.

2

HARVEY WILEY

1844–1930

Field | Chemistry
Major Contribution | Crusader for pure foods and father of the Food and Drug Administration
Indiana Connection | Born and raised in Kent, attended Hanover College and Indiana Medical College, professor at Purdue

We sit at a table delightfully spread,
And teeming with good things to eat,
And daintily finger the cream-tinted bread,
Just needing to make it complete
A film of the butter so yellow and sweet,
Well suited to make every minute
A dread of delight. And yet while we eat
We cannot help asking "What's in it?
Oh, maybe this bread contains alum and chalk,
Or sawdust chopped up very fine,
Or gypsum in powder about which they talk,
Terra alba just out of the mine.
And our faith in the butter is apt to be weak,
For we haven't a good place to pin it
Annato's so yellow and beef fat so sleek,
Oh, I wish I could know what is in it?

—section of a poem penned by Harvey Wiley

HARVEY WASHINGTON WILEY was born in a log cabin on a farm near the small town of Kent, in Jefferson County, Indiana, on October 18, 1844. He was the sixth of seven children—four boys and three girls—born to Preston Pritchard Wiley and Lucinda Weir Maxwell. In his autobiography, Wiley says his father "had practically no advantages of education, even of the common school, but was a man of keen intellect, fine judgement and high powers of reasoning. He succeeded admirably in educating himself." Of his mother, he said that although she had a total of only three months in school, "Her natural wit was well known throughout the whole neighborhood, and her keenness of observation and intellectual perception were recognized by all who knew her."

Like most pioneer families, the Wileys were extremely religious. Sundays were reserved exclusively for church-going and Bible-reading. Wiley recalls innocently building sand piles near a creek one Sunday and receiving two blows with a switch for the sin. The punishment was administered the following Monday, since nothing could be done on a Sunday. The Wileys coupled their religious fundamentalism with social progressivism. They were firmly against slavery, and the family spent many evenings reading aloud from the abolitionist novel *Uncle Tom's Cabin*. Preston believed in acting on one's beliefs, a trait that he inculcated into his son, and became a conductor on the Underground Railroad. He met people escaping from slavery at night on the banks of the Ohio River and escorted them ten miles to the village of Lancaster, where they hid during the day and traveled to the next stop at night.

Wiley spent much of his time helping out on the farm. As a child, he collected firewood, carried spring water and herded cattle. When he was older, he plowed fields, planted crops and helped with the harvest. One year, Preston received a few packages of sugarcane seed and gave them to Wiley, who planted the seeds and harvested the cane in the fall. Wiley vividly recalled placing a piece of cane in his mouth and tasting the sweetness. Impressed with the success of the crop, Preston helped his son extract the juice and make sorghum syrup. This experience piqued Wiley's interest in sugar and food processing.

Wiley's early education was sporadic. At the time, there was no free public education. Instead, schools operated on a subscription basis, with each student paying a small fee. Wiley attended a subscription school whenever he could. At age eighteen, Wiley told his father he was going to Hanover College, just a few miles down the road. Preston said simply, "Very well, son, go along!" On April 9, 1863, Wiley put on his homespun suit, walked five miles to the school and enrolled.

Wiley's time at Hanover was interrupted by service in the Civil War. He was assigned the duty of guarding a Union army supply depot in Tennessee, where he contracted the measles and suffered the symptoms, including chronic diarrhea, for months. When he was finally discharged in October 1864, he described himself as "more dead than alive," and his weight on his large six-foot one-inch frame was down to 119 pounds. Back home, his mother nursed him back to health through the long winter, and by the end of April 1865, he was able to resume his studies at Hanover.

Wiley graduated in 1867, decided to become a physician and taught school for a few months to save up enough money for medical school. He began his medical training in the summer of 1868 with an apprenticeship under Dr. S.E. Hampton, whom he had met during the Civil War. Hampton was a country doctor in Milton, Kentucky, just across the Ohio River from Madison, Indiana. Wiley studied Hampton's medical books, and the doctor quizzed him on the contents. The most valuable part of the experience was riding out with Hampton to see patients. Wiley later claimed that "making the rounds with Doctor Hampton gave me better ideas of practical medicine than I could possibly have got out of books."

In the fall of 1868, he continued his studies at the Indiana Medical College in Indianapolis, a branch of Indiana University. He earned his medical degree in the spring of 1871 and, hoping to save some money before starting his medical career, accepted a position as a science teacher at Central High School in Indianapolis. The next spring, he was unexpectedly invited to teach chemistry at his alma mater, the Indiana Medical School. Wiley agreed to take the job but, feeling unprepared, asked to be allowed to study advanced chemistry before teaching the subject.

The school agreed, and Wiley wound up at the most prestigious college in the country: Harvard University. In April 1873, Wiley's professor at Harvard suggested that he apply for a degree. This involved taking a long series of exams, both written and oral, lasting seventeen days. He passed the exams with ease, prompting him to later proclaim that he "was perhaps the most rapidly catapulted graduate the Lawrence School ever had." Thus it was that Wiley received his bachelor of science degree in chemistry from Harvard University in about five months, an occasion that Wiley later designated as the "proudest day of my life."

Upon returning to Indianapolis, Wiley postponed his medical practice for a year while he held down three jobs: teaching chemistry classes at what is now Butler University in the morning, at the city high school in the afternoon and at the Medical College in the evening. His combined

Harvey Wiley portrait. *Shutterstock.*

income from the three jobs allowed him to buy a house in Irvington, three miles outside of Indianapolis, and he invited his parents to move in with him. But three jobs was too much. One night in February, he developed severe pains throughout his body that almost paralyzed him. The doctors diagnosis was spinal meningitis. He was unconscious for three weeks and developed a swelling on his right leg below the knee. The doctors were about to amputate the limb when Wiley requested that they let him "die whole." Instead, the doctor operated on the swollen area, and his condition gradually improved. He lost his hair but regained his strength and fully recovered in a few months.

In August 1874, Wiley got yet another surprise. He was invited to join the faculty as a professor of chemistry at the newly established Purdue University in West Lafayette at a salary of $2,000 per year with free housing in the dorms. The job offer came courtesy of A.C. Shortridge, a former superintendent of schools in Indianapolis who was now the new president of Purdue. Shortridge knew Wiley through his teaching at the high school. Wiley, the youngest man on the Purdue faculty, taught physics and chemistry classes at Purdue and was popular with students. He was the business manager of the university's first newspaper, founded the school's first military cadet program (the "Purdue Army") and organized the first baseball and football teams.

In the spring of 1878, Wiley took an unpaid leave of absence from Purdue to travel and study in Europe. He crisscrossed the continent attending lectures, visiting hospitals and learning how to use the latest scientific equipment. Wiley returned to Purdue with renewed enthusiasm for the teaching profession and decided to abandon his plan to become a practicing physician. He also came back with specialized instruments for analyzing food chemistry, paid for out of his own pocket because the university declined payment. Wiley used the equipment to study sugar and sorghum chemistry, with the goal of helping the sugar industry in the United States. The cost of importing sugar was high, and the country needed to find a domestic source. Wiley made presentations to the American Association for the Advancement of Science and developed a reputation as one of the nation's leading sugar chemists.

Wiley enjoyed a good relationship with the Purdue administration until 1880, when he bought a high-wheeled bicycle and had the audacity to ride it around the campus wearing knee breeches. This was evidently too much for the board of trustees, who summoned Wiley in for a little chat. As he recalled in his autobiography, Wiley thought his salary was going to be raised when one of the trustees stood up and made the following statement:

The disagreeable duty has been assigned to me to tell Professor Wiley the cause of his appearance before us. We have been greatly pleased with the excellence of his instruction and are pleased with the popularity he enjoys among his pupils. We are deeply grieved, however, at his conduct. He has put on a uniform and played baseball with the boys, much to the discredit of the dignity of a professor. But the gravest offense of all has lately come to our attention. Professor Wiley has bought a bicycle. Imagine my feelings and those of the other members of the board on seeing one of our professors dressed up like a monkey and astride a cartwheel riding along our streets. Imagine my feelings when some astonished observer says to me, "Who is that?" and I am compelled to say, "He is a professor in our university!" It is with the greatest pain that I feel it is my duty to make these statements in his presence and before this board.

Amused, but frustrated, Wiley responded: "Gentlemen, I am extremely sorry that my conduct has met with your disapproval. I desire to relieve you of all embarrassment on these points. If you will give me pen and paper I shall proceed to do so." Wiley wrote out a letter of resignation, gave it to the secretary and left. The next day, Wiley received a note from the board of trustees saying they refused to accept his resignation.

In 1881, Wiley was asked by the Indiana State Board of Health to inspect syrups, sugars and honey on sale in the state and test them for purity. He found that nine out of ten syrups tested were out-and-out fakes. Bottles labeled as maple syrup were actually filled with corn syrup. Honey was also often counterfeit—even the chunk of honeycomb in the jar was actually made of wax. He found impurities in the food, including copper, sulfuric acid and ground-up animal bones. That summer, Wiley's eye-opening report was published in the state record and in *Popular Science* magazine. It made a few people happy—makers of real maple syrup were delighted. But it irritated many others—corn farmers, corn syrup makers and mislabeled product bottlers expressed their displeasure. Undaunted by the criticism, Wiley wrote a second report to the state board recommending that Indiana lawmakers establish food purity standards. It was Wiley's first foray into the pure foods issue.

In December 1882, Wiley gave a speech at a meeting of sugarcane growers in St. Louis. In the audience was George Loring, head of the United States Department of Agriculture (USDA). Loring was looking to replace Peter Collier, the department's current chief chemist, who had complained about Loring to the Washington press, resulting in newspaper stories that reflected poorly on Loring. Wiley impressed Loring, and several months later, Loring offered him the job.

The offer came at just the right time. Wiley was increasingly frustrated by the Purdue administration. In his autobiography, he explains, "I honestly felt that my sphere of influence at the school was too restricted. What I considered academic freedom was an unknown quantity under the administration at that time. There were many red-tape regulations enjoined upon teachers and students which were very distasteful to me." The university president had also opposed the presence of Greek fraternities, which Wiley supported, although he was generally against secret societies. And of course, there was the bicycle kerfuffle, from which Wiley was still "smarting painfully."

So after nine years at Purdue, Wiley resigned his position and took the oath of office as chief of the USDA's Division of Chemistry on April 9, 1883. To understand what Wiley accomplished there requires some historical context. By the mid-nineteenth century, foods sold in the United States had an often dubious and sometimes even dangerous reputation, largely due to fillers added to make the product go further. Ground-up insects were added to brown sugar, scorched sawdust and seeds were mixed in with coffee, and spices were contaminated with pulverized coconut shells, charred rope and floor sweepings. Milk was particularly bad. In addition to being diluted by pouring in a pint of water to every quart of milk, plaster of Paris or chalk was sometimes added as a whitener, with the concoction occasionally topped off with pureed calf brains to provide a layer of cream.

By the late 1800s, the food industry had also begun adding preservatives to make products last longer. One of the most popular preservatives was formaldehyde, used to extend the shelf life of milk and meat in a time before refrigeration. Other common preservatives included borax, a mineral also used as a cleaner, and salicylic acid, a chemical used in certain drugs. Not only did all these additives and preservatives cheat customers out of product, but also, and more importantly, they were a public health hazard. People were getting sick and even dying from the food they ate. At the time, the only government agency responsible for chemically analyzing food was the

Harvey Wiley (*standing in front*) with some of his staff at the Division of Chemistry during the 1890s. *Flickr Commons.*

Department of Agriculture, created by Abraham Lincoln in 1862. And now the USDA had Harvey Wiley, the right man for the job.

Wiley moved to Washington, D.C., rented a bedroom in a family home where he stayed for the next twenty years and went to work. Wiley figured that if he could prove food adulteration went beyond mere cheating to actually being harmful to human health, the public and Congress would support a national food safety policy. So in 1901, he asked Congress to fund a series of studies that he called "hygienic table trials" in which human test subjects would sit down at a clean (hygienic) table and eat carefully controlled and measured meals. Some of the subjects would eat food containing the preservatives while the others ate food that was preservative free. The idea was to determine which preservatives, if any, were unsafe for human consumption and at what dosage. The next year, Congress gave Wiley a grant of $5,000 ($150,000 in today's dollars). It was only a third of Wiley's requested amount, but it was a start.

As described in Deborah Blum's excellent book *The Poison Squad*, Wiley assembled a team of a dozen healthy young male volunteers, mainly from the Department of Agriculture, to participate in the study. The makeshift dining room and kitchen were housed in the basement of the USDA building. The men were fully aware that they were consuming potential poisons, but they did not know which foods actually contained the chemicals. The volunteers agreed to participate for six months and to not hold the government responsible for any resulting injury or illness. They pledged to eat all their meals at the "hygienic table" and not consume any outside food or drink except water. In addition, each participant recorded his weight, pulse and body temperature before each meal and had to submit daily urine and stool samples for analysis. The studies also looked at whether the preservatives were eliminated from the body by respiration or perspiration. Physicians closely monitored the health of the volunteers and recorded any symptoms. Their only rewards, besides contributing to scientific knowledge, were free meals and five dollars a month.

The research began in November 1902 and quickly became a national sensation. One reporter dubbed the brave volunteers the "Poison Squad," and Wiley earned the moniker "Old Borax." Although he had a good sense of humor, Wiley worried that all the lighthearted press coverage might discredit the serious nature of the work. At one point, Wiley discovered that a reporter had been chatting with the chef through a basement window, so he warned his employees not to give interviews and told volunteers they would be dismissed from the study if they talked to the press.

As a result of the research, the momentum for federal laws regulating foods and drugs started to build. Although women couldn't vote, they could and did apply political pressure. One million women wrote the White House in favor of food safety regulation. The cause was also aided by the publication of *The Jungle*, a novel by Upton Sinclair. Sinclair had gone undercover for

Harvey Wiley (*standing*) with the "Poison Squad." *Wikimedia Commons.*

seven weeks in the Chicago stockyards. Based on his experiences, the book was an exposé on the working conditions in the Chicago meatpacking industry. Sinclair had intended to advance the cause of socialism, but most readers were alarmed by the unsanitary conditions. Of the public reaction, Sinclair said with a twinge of disappointment, "I aimed at the public's heart, and by accident I hit it in the stomach." One person who read the book was President Theodore Roosevelt, who invited Sinclair to the White House to discuss the issue.

Roosevelt did not trust the socialist Sinclair, so he sent a pair of independent investigators, Commissioner of Labor Charles P. Neill and social worker James B. Reynolds, to Chicago to look into the matter. What they found in Chicago was even worse than the conditions portrayed in the book. Here is a small sample from what became known as the Neill-Reynolds report: "In a word, we saw meat shoveled from filthy wooden floors, piled on tables rarely washed, pushed from room to room in rotten box carts, in all of which processes it was in the way of gathering dirt, splinters, floor filth and expectoration of tuberculous and other diseased workers." Although convinced that the industry needed reforming, Roosevelt decided against publishing the report and instead showed it to a few congressmen to gain political leverage. Still, the legislation got nowhere.

Sinclair had had enough. On friendly terms with Neill and Reynolds, he learned what he could about the report, stuffed relevant documents into a briefcase and walked into the offices of the *New York Times*. The explosive story appeared on the front page of the newspaper on Monday, May 28, 1906. In the story, Sinclair spoke of seeing "with my own eyes the doctoring of hams that were so putrefied that I could not force myself to remain near them." Neill recalled that "the pillars of the buildings were caked with flesh" and that "the meat is dragged about on the floor, spat upon and walked upon."

Roosevelt then agreed to release an eight-page summary of the Neill-Reynolds report; it was published verbatim in newspapers across the country. People were appalled. European nations vowed to stop importing American meat. A new avalanche of letters flowed from women's clubs. The American Medical Association sent telegrams. And Wiley presented new evidence from the Poison Squad research. The battle was finally won on June 30, 1906, when President Theodore Roosevelt signed into law both the Meat Inspection Act and the Pure Foods and Drugs Act. The latter, called "Dr. Wiley's Law" by the newspapers, was the first federal law regulating the development and production of safe foods and drugs.

With the passage of the law, Wiley's tenure at the bureau became increasingly controversial. He had enemies in Congress, in the food industry and even within his own department. Specifically, Wiley and Secretary of Agriculture James Wilson began having conflicts over policy. Eventually, Wilson appointed a panel of scientists to review the Bureau of Chemistry's work. Wiley's opinion was that the group, known as the Remsen Board after chairman Ira Remsen, "was the cause of nearly all the woes that subsequently befell the Pure Food Law." The *New York Times* agreed, bluntly writing in an editorial that the board "was created on February 20, 1908, for the specific purpose of overruling the findings of Dr. Harvey Wiley of the Bureau of Chemistry with respect to the purity of food and drugs."

More controversy erupted when Wiley decided to pick a fight with Coca-Cola. By the beginning of the twentieth century, Coca-Cola had become a popular soft drink, especially with prohibitionists, who saw it as an alternative to alcohol. The name of the drink came from two of its original ingredients: coca leaves and kola nuts. The cocaine from the coca leaves had, by that time, been removed. Any remaining stimulant value from the drink came from caffeine, an ingredient that Wiley thought was potentially dangerous, especially to young children who consumed the sugary beverage. The chairman of the company, Asa Chandler, insisted that the drink was safe, but that didn't satisfy Wiley. On October 9, 1909, Wiley had the government seize a shipment of Coca-Cola as it was being transported from the main plant in Atlanta to the bottling plant in Chattanooga and had it declared contraband under the Pure Food and Drug Act. Thus began one of the most memorable court battles involving food products: *United States v. Forty Barrels and Twenty Kegs of Coca-Cola.*

The trial was quite a spectacle and made newspaper headlines across the country. On one side was Wiley, backed by the U.S. government; on the other side was Asa Chandler and one of the largest corporations in America. The government claimed that Coca-Cola was adulterated and misbranded. The adulteration was from the caffeine, which might be hazardous to human health. The misbranding was based on a chemical analysis showing the product contained no coca and little if any kola. According to the trial transcript, the actual ingredients included water, sugar, caffeine, glycerin, lime juice and other flavoring chemicals. The company responded by saying that the formula had been followed for more than twenty years without any deleterious effects and that Coca-Cola was simply a brand name rather than a descriptive label. The trial began on March 13, 1911, in a federal courtroom in Chattanooga, with each side lining up a parade of witnesses. Some of the testimony was laughable. Religious fundamentalists, for example, testified that drinking Coca-Cola caused its consumers to engage in wild partying, led to sexual indiscretions by coeds and caused boys to masturbate. In the end, after months of testimony and appeals, Coca-Cola reached a settlement with the government and agreed to reduce the level of caffeine by half.

At about the same time as the war over Coca-Cola, Wiley's conflicts with his boss, James Wilson, came to a head. Wilson tried to get Wiley fired over a technicality: an expert working in his department had been paid in excess of the legal rate. The charge was investigated and resulted in a letter from President William Howard Taft saying that he had found no evidence of any wrongdoing in the payment arrangement, thus fully exonerating Wiley of any charges. A congressional committee also looked into the scandal and concurred with the president. Wiley issued a statement thanking the president for his sense of fairness, the press for their near unanimous support and the many people who had written him letters of encouragement.

Although Wiley had been cleared, Wilson was still secretary of agriculture, and the antagonism between the two men continued. Wiley's recommendations were routinely ignored, and the food industry was bypassing him, taking complaints directly to Wilson. Not wishing to remain in such an openly hostile environment, Wiley resigned from his position as head of the Bureau of Chemistry on March 15, 1912. The public reaction was one of sadness, with one newspaper headline reading: "Women Weep as Watchdog of the Kitchen Quits After 29 Years."

Not all of 1911 was bad for Wiley. A bachelor all his life, Wiley married Anna "Nan" Campbell Kelton. It was a May-December romance: he was sixty-six, she was thirty-four. The two had first met in 1896 when she was a USDA librarian. It was love at first sight for Wiley, who recalled, "I saw a young woman with a book in her hand, apparently looking for the proper place to deposit it. I was immediately struck by her appearance." Wiley claimed that he grabbed the arm of Edward Cutter, the library manager, and after asking for the woman's name, said, "Cutter, I intend to marry that girl." Cutter wittily replied, "Perhaps it would be well for you to meet this young lady before proposing matrimony." Over the next several years, Wiley courted Kelton, and the couple attended plays and concerts. Wiley proposed to Kelton in the spring of 1900, but because of the age difference, her desire to remain independent and her mother's disapproving attitude toward the relationship, she declined. But Wiley kept her photo in the cover of his pocket watch.

Ten years later, Wiley was waiting for a streetcar when he saw her again. Emboldened by her warm response, he asked if he could take her to dinner or a show. Wiley was surprised when she said yes. The couple quickly rekindled their romance, and a few weeks later, Wiley again proposed marriage. This time, she accepted. The engagement made newspaper headlines across the country. The *Los Angeles Examiner* ran the headline "Greatest Enemy of Food Germs Falls Victim of Love Microbe," while the *Chicago Tribune* went with "Pure Food Expert to Desert Cafes for Home Cooked Meals." A cartoon in the *Denver Post* showed Kelton fainting as her husband inspected the kitchen for additives and preservatives. The happy couple got married and, much to their surprise, were soon expecting their first of two sons.

Wiley spent the rest of his life at *Good Housekeeping* magazine, where he continued his fight for pure foods. In 1914, Wiley coauthored an exposé on obesity treatments titled "Swindled Getting Slim." The article described how diet charlatans sold misleading and fraudulent products without interference from government regulators. In 1921, Wiley's articles helped pass the federal Maternity Bill, which provided funding for improved infant care and led to a reduction in the infant mortality rate. And as early as 1927, Wiley wrote that the use of tobacco might be harmful and may even be a cause of cancer. As the years passed, evidence supporting Wiley's suspicion mounted. In 1952, *Good Housekeeping* stopped accepting cigarette advertisements. Twelve years later, the surgeon general put out a report warning of the dangers of smoking.

Harvey Wiley died at his home in Washington, D.C., on June 30, 1930, the twenty-fourth anniversary of the signing of the Pure Foods and Drugs Act. He was buried in Arlington National Cemetery with full military honors because of his service in the Civil War. Anna asked the minister to base his tribute on a scripture from Second Timothy that reads "I have fought the good fight, I have finished my course, I have kept the faith." His tombstone reads "Father of the Pure Food Law." Wiley's Bureau of Chemistry would eventually become the U.S. Food and Drug Administration (FDA).

3

VESTO SLIPHER

1875–1969

Field | Astronomy
Major Accomplishment | Discovered that nearly all galaxies are moving away from our own Milky Way Galaxy
Indiana Connection | Born in Mulberry, went to high school in Frankfort and attended Indiana University, where he earned a doctoral degree in 1909

> *Someone else may have accomplished much more, but surely no one could find more pleasure in doing it than I.*
> *—Slipher in a speech to the National Academy of Science on the occasion of being awarded the Henry Draper Medal*

VESTO MELVIN SLIPHER (usually referred to as V.M.) was born on a farm in the small town of Mulberry, Indiana, near Lafayette, on November 11, 1875, to parents Hannah App and Daniel Clark. Vesto had a younger brother, Earl C. Slipher, born in 1883, who also became an astronomer. Not much is known about Slipher's childhood, but the physical demands of life on the farm probably helped prepare Slipher for the rigors of a career in observational astronomy in the early 1900s. In fact, Slipher emphasized the importance of physical health to young men considering a career in astronomy.

Slipher graduated from high school in Frankfort, Indiana, and taught for a few years at a little country school north of town. At the age of twenty-

Vesto Slipher in his Indiana University yearbook photo. Slipher is standing in the back row, second from the left. *Flickr Commons.*

one, he decided to go to college and began his studies at Indiana University in 1897. There, a professor ignited his interest in astronomy, and he earned a degree in the subject in 1901. After graduation, Slipher landed a job as a temporary assistant at the Lowell Observatory in Flagstaff, Arizona. The temporary position lasted over half a century.

The Lowell Observatory was founded and funded by Percival Lowell, a rich Bostonian and avid amateur astronomer. Lowell developed a keen interest in the planet Mars and was convinced that it was home to an ancient and advanced civilization. He had an observatory built on a hill in Flagstaff so he could study the red planet. When Lowell looked at Mars through the telescope, he thought he saw canals that radiated outward from central points. He drew intricate maps of the canal system and wrote three books about Mars that captured the public's imagination. We now know that the canals Lowell claimed to have seen weren't really there at all but were just products of his own wishful thinking. What did Slipher think about Lowell's Martians? Slipher was more cautious, but he believed that intelligent life on Mars was certainly a possibility.

Slipher arrived at the Lowell Observatory on August 10, 1901. Shortly thereafter, a state-of-the-art spectroscope was delivered to the observatory. A spectroscope is an instrument that dissects light into its constituent wavelengths much like a prism spreads light into its rainbow of colors. Slipher's first job was to mount the spectroscope on the big two-foot-diameter refracting telescope, make any necessary adjustments and learn how to use it. By mid-1902, Slipher had produced spectrograms of Mars, Jupiter and Saturn and used them to confirm the known periods of rotation of those planets. (The period of rotation is a day—the time it takes a planet to spin once on its axis.) He went on to determine the rotation periods for several other planets. Slipher was more interested in the stars than the planets, but Lowell had a policy that planetary research must have first priority. Lowell wrote Slipher a note in 1904 saying, "With regard to yourself, by all means make your star measures for velocity—whenever there is no pressing planetary work—and good luck to you in the result." Slipher adhered to the policy and pursued his own interests as time permitted. His first scientific paper, published in the *Astrophysical Journal* in 1902, was a report on the speed of the star Zeta Herculis.

On New Year's Day 1904, back in Frankfort, Slipher married Emma Rosalie Munger. They returned to Flagstaff, bought a home on the grounds

The twenty-four-inch Clark refracting telescope used by Vesto Slipher at Lowell Observatory in Flagstaff, Arizona. *Shutterstock.*

of the observatory and had two children. In 1908, Slipher's brother Earl joined him at the observatory. Earl would go on to become a noted planetary astronomer and a pioneer in planetary photography, taking nearly 200,000 images of Mars. Earl also got involved in politics, serving as the mayor of Flagstaff and in the Arizona state legislature.

In 1909, Slipher earned his doctoral degree in astronomy from Indiana University. Over the next several years, he discovered that gas and dust exists in interstellar space. This result brought congratulatory letters from several prominent astronomers, but for nearly twenty years the discovery was largely ignored by the astronomical community.

Slipher's next discovery could not be ignored. There were objects in the night sky called spiral nebula, but at the time, nobody really knew what they were. They were faint, diffuse and numerous. Astronomers had spent three hundred years speculating about them. A few thought they were baby solar systems in the process of formation. Some suggested that they were dense clusters of stars within our own Milky Way Galaxy. Others postulated they were "island universes," vast conglomerations of stars lying beyond the confines of our own Milky Way Galaxy. Slipher's discovery would play an important role in solving the mystery of the spiral nebulae.

In the fall and winter of 1912, Slipher took a set of four spectrographs of the Andromeda nebula that were detailed enough to allow him to measure its radial velocity (the speed with which an object moves toward or away from the earth). The radial velocities of about 1,200 stars had been measured to be a few tens of kilometers per second (km/sec). No one had any reason to believe that the velocities of the nebulae would be any different. After checking his calculations, Slipher wrote to Lowell on February 3, 1913, and told him that the Andromeda nebula was approaching the earth at a speed of 300 km/sec (190 miles/sec), much faster than the stars. Lowell wrote back to Slipher, "It looks as if you have made a great discovery. Try some other spiral nebulae for confirmation."

Slipher complied and made a similar measurement for other spiral nebulae. At a meeting of the American Astronomical Society in August 1914, Slipher announced the radial velocities of fifteen nebulae and reached a remarkable conclusion. "In the great majority of cases, the nebula is receding," he told his fellow astronomers. "The largest velocities are all positive....The striking preponderance of the positive sign indicates a general fleeing from us or the Milky Way." After finishing his talk, Slipher received a rare standing ovation. In 1917, the Dutch astronomer Willem de Sitter first suggested that the universe might be expanding. By

Slipher found that this galaxy, the nearby Andromeda Galaxy, is moving toward our own Milky Way galaxy. Galaxies outside our local group are all moving away. *Shutterstock.*

then, Slipher's list of spiral nebular velocities had grown to twenty-five and remained "preponderantly positive"—that is, moving away. By 1922, his list had expanded to forty-one velocities.

In 1929, Edwin Hubble measured the distances to the nebulae on Slipher's list and, as he later wrote to Slipher, using "your velocities and my distances" discovered a simple direct relationship between velocity and distance: the farther away a nebula was, the faster it was moving away. Twice as far away meant that the nebula was moving away twice as fast. We now know the "nebulae" are actually galaxies, and the relationship became known as Hubble's Law. (Hubble also has a brief Indiana connection, having taught and coached basketball at New Albany High School during the 1913–14 school year.)

Also in 1929, Slipher hired a young assistant named Clyde Tombaugh, to whom he assigned the task of searching for Planet X, a hypothesized planet beyond the orbit of Neptune. Percival Lowell predicted the existence of Planet X in 1902 based on tiny deviations in the expected orbit of Uranus. In January 1930, Tombaugh discovered Planet X, and in May, Slipher proposed that the new planet be named Pluto, after the Roman god of the underworld. The name, suggested by an English schoolgirl, seemed appropriate for an object orbiting on the edge of the solar system. And the first two letters of Pluto just happened to be Percival Lowell's initials, a fitting tribute to the man who predicted the planet's existence. Pluto's tenure as a planet lasted seventy-six years; in 2006, the International Astronomical Union demoted Pluto to dwarf planet status.

Slipher's physical vitality continued into his later years. A friend recalled how in his sixties Slipher could easily climb the twelve-thousand-foot San Francisco Peaks north of Flagstaff. Slipher "was always ahead of us 'boys' climbing the mountain—we puffing and panting and he, disgusted, waiting for us to catch up." A fellow astronomer remembers that "V. M. at age sixty-five could chop wood with the best of them." Slipher retired from the Lowell Observatory in 1952 at the age of seventy-seven. He died in 1969, just a few days before his ninety-fourth birthday, and is buried in Citizens Cemetery in Flagstaff.

4

HERMANN MULLER

1890-1967

Field | Biology
Major Contribution | Won the Nobel Prize in 1946 for Medicine or
    Physiology for discovering that X-rays cause genetic mutations
Indiana Connection | Professor of zoology at Indiana University from
    1945 to 1964

> *The central problem of biological evolution is the nature*
> *of mutation, but hitherto the occurrence of this has been*
> *wholly refractory and impossible to influence by artificial*
> *means, although a control of it might obviously place the*
> *process of evolution in our hands.*
> —H.J. Muller, quoted in Genes, Radiation, and Society:
>   The Life and Work of H.J. Muller, *by Elof Axel Carlson,*
>   *1981, 104*

HERMANN JOSEPH MULLER was born in New York City on December
20, 1890. He was the son of Hermann Joseph Muller Sr., part-owner of a
metal art business, and Frances Lyons. Muller was raised in the Unitarian
Church, but by high school, he was an atheist. His parents instilled in their
son a love of nature and a concern for the welfare of people, as evidenced
by their support of the emerging labor and socialist movements. Muller was
only ten years old when his father died from a stroke. After that, the family,
which included a sister named Ada, lived on a modest income from the
metal art business partnership.

Hermann grew up in Harlem, where he attended elementary school, and went to Morris High School in the Bronx. He excelled in school and was awarded a scholarship to Columbia University at age sixteen based on his entrance exam scores. Muller worked his way through school and, by the end of his freshman year, was fascinated by the field of biology. He read a book on genetics and took biology classes, where he learned about the exciting microscopic world of cells and chromosomes.

After graduating in 1910, Muller decided to pursue a doctorate at Columbia. He worked under eventual Nobel Prize–winning biologist Thomas Hunt Morgan, whose laboratory was known as the "fly lab" because it used fruit flies (the scientific genus name is *Drosophila*) to study genetics. Fruit flies are well suited for genetics research because they are cheap to raise, reproduce prolifically in small bottles and have a breeding time of about ten days, so that multigenerational experiments can be performed quickly. Working with Morgan and fellow students Alfred Henry Sturtevant and Calvin Blackman Bridges, Muller helped establish the science of classical genetics. Their work showed that genetic mutation was the mechanism behind the evolution of life by natural selection. Muller's contributions to the fly group were primarily theoretical—he made predictions for the outcomes of new experiments and provided explanations for experimental results.

Vial containing fruit flies (*Drosophila melanogaster*). No other animal has contributed as much to the field of genetics. *Shutterstock.*

In 1914, the fly lab was visited by Julian Huxley, who had been appointed chair of the biology department at the recently founded Rice Institute (now Rice University) in Houston. Huxley was impressed by what he saw and asked Morgan to recommend a student for a position in his new department. Morgan recommended Muller. Eager to take the job, Muller rushed to finish writing his dissertation on a process involving chromosomes known as "crossing over." By the beginning of the 1915–16 academic year, Muller had moved to Texas.

At Rice, Muller taught biology and continued his work with fruit flies. He made a lifelong friend in Huxley, and both men supported the hypothesis of a genetic role in natural selection. In 1917, Huxley, an Englishman, felt obligated to return to England to enlist in the war effort and left Muller in charge of the biology department.

In 1919, Morgan went on sabbatical leave, and Muller was asked to take his place temporarily at Columbia. He accepted the invitation, moved back to New York City and continued his research. Muller had hoped to join the Columbia faculty on a permanent basis, but the department chair worried that things wouldn't work out once Morgan came back. In 1921, Muller returned to Texas, but not to Rice. Instead, he accepted a position at the University of Texas at Austin.

Muller's tenure at Texas was marked by personal turmoil and professional triumph. The turmoil started when two of his colleagues, influenced by Muller's work with fruit flies, decided to switch to Drosophila in their own research. This led to a rivalry between the three scientists, with Muller convinced that his colleagues were stealing his ideas. Muller had soon alienated the two and began working at night to avoid them. But this led to more problems because Muller had recently married Jessie Marie Jacobs, a mathematics professor. Working at night did not sit well with his new bride, and the couple began to have marital difficulties.

Muller was also sympathetic to the Communist Party, although he never officially joined. It is important to keep in mind the historical context—this was just after the Russian revolution, when many people, especially academics, had a rather rosy and optimistic view of communism. He became an advisor to the National Student League, which the FBI viewed as a communist student organization, and became an underground editor for *The Spark*, a newspaper, still in existence today, that supported progressive ideas such as civil rights, women's rights, unemployment insurance and social security.

In spite of the chaos in his personal life, Muller did some of his best scientific work during this period. By 1925, he had: identified the gene as the basis of life since all other cellular components come from it; argued that the first living things were gene-like molecules that could copy themselves along with any new variations; worked out the first mutation rates and showed they could vary within a species; and showed that genes could be changed by modifier genes or by changes in the environment such as temperature. Beginning in 1926, Muller made a series of major discoveries. In November, he performed experiments exposing genes to varying doses of X-rays and found a clear connection between radiation dose and lethal and visible mutations. This launched a new area of science called radiation genetics. In 1927, Muller published his results in the journal *Science* and gave a lecture at the International Congress of Genetics in Berlin. The publicity surrounding Muller's Berlin report about artificially induced mutations gave him international stature and made him one of the best-known intellectuals of the time. By 1928, other scientists had replicated his results and expanded them to other organisms like wasps and corn. In the years that followed, Muller publicized the probable dangers of exposure to radiation in people who operate X-ray equipment, including doctors and shoe salesmen who radiated customers' feet.

Meanwhile, Muller's personal life continued to deteriorate. His marriage was in trouble, the FBI was investigating and the local newspapers contained veiled references to him as a communist subversive. To make matters worse, Muller had raised eyebrows by inviting two Soviet students to work in his lab. One day, Muller disappeared. He didn't show up for his classes, he wasn't in the lab and he wasn't at home. An anxious call from his worried wife prompted a search party to go out looking for him. They found him in the woods on the outskirts of Austin walking around in a confused state with his clothes wrinkled and muddy. It turned out that he had attempted suicide by taking an overdose of barbiturates. He returned to class the next day as if nothing had happened.

Immediately after the suicide attempt, Muller left to make two important presentations. The first was to the Third International Congress of Eugenics at the American Museum of Natural History in New York City. Muller had been interested in the topic of eugenics, the idea that the human race can be improved by selectively breeding people with desirable hereditary traits, ever since his college days. He published a paper in 1925 on identical twins raised apart in which he argued that little was known about the genes involved in human behavior and about how behavior could be affected by the environment.

At the eugenics conference, in spite of attempts to stop his presentation, Muller let loose with a full-throated denunciation of the American eugenics movement. He argued the idea that negative personal traits such as criminality, feeblemindedness, alcoholism and vagrancy were largely innate, was unproven and probably false. He suggested that eugenics might possibly succeed in a socialist country, where all children, regardless of race or gender, had an equal opportunity for education, housing and other social services. He concluded with an impassioned appeal for racial equality, equal rights for women and an end to claims of superiority based on socioeconomic class. Muller's speech made newspaper headlines around the world. His next stop was the Sixth International Congress of Genetics at Cornell University in Ithaca, New York, where he presented a lengthy paper titled "Further Studies on the Nature of Gene Mutation" in which he put forth a theory about gene function. Many who heard Muller's talk were stunned by its originality and importance.

In September 1932, Muller packed his bags for what was supposed to be a one-year European sabbatical but turned into an eight-year odyssey encompassing four countries. The first stop was Berlin, where Muller had been awarded a Guggenheim Fellowship to study at the Institute for Brain Research, part of the prestigious Kaiser Wilhelm institutes. But Adolf Hitler soon came to power, and the Nazis, suspecting communist activity, vandalized the institute. This prompted Muller to accept an invitation from Nikolai Vavilov, the Soviet equivalent of the American secretary of agriculture, to come to the Soviet Union and establish a genetics laboratory in Leningrad.

Muller spent the next five years in the Soviet Union. He briefly reconciled with his wife, and she and their son David moved to Leningrad to be with Hermann. As a corresponding member of the Soviet Academy of Sciences, Muller could count on generous financial support. He put the money to use and purchased the necessary equipment for a fruit fly lab, recruited graduate students, hired research associates and set up shop as the Institute of Genetics. As an added bonus, there were no teaching duties. The research projects he initiated focused on gene function, structure and evolution. In 1934, the institute moved to Moscow. The next year, Muller and his wife divorced, and she and David moved back to America.

Once again, Muller turned his attention to eugenics. Although he had blasted the eugenics movement in the United States, he thought it might actually work in a socialist country. He did research using identical twins, studying their physical traits, performance on tasks and susceptibility to disease, to try to sort out genetic factors from environmental influences. He finished a book on eugenics, titled *Out of the Night*, that he had started in 1919. When he asked a Communist Party member how to begin a eugenics program in the Soviet Union, he was advised to start at the top. So Muller had the book translated into Russian, wrote a long letter explaining his ideas about eugenics and forwarded the book and letter to Joseph Stalin.

Unfortunately for Muller, an alternate view of heredity was taking hold in the Soviet Union. This movement, led by agronomist Trofim Lysenko and known as Lysenkoism, rejected the concept of the gene, Mendelian heredity and evolution by natural selection. Proponents claimed that rye could be transformed into wheat and wheat into barley, that weeds could spontaneously transmute into grain and that "natural cooperation" rather than natural selection was the driving force behind evolution. They dismissed Western genetics as a capitalist, bourgeois system that would hinder progress and promote fascism.

Lysenkoism was a political philosophy masquerading as science; it was completely and absolutely wrong. But Stalin supported the movement, which effectively ended any argument about it. Over three thousand mainstream Soviet biologists were fired, imprisoned or executed. Vavilov, who had invited Muller to the USSR, was arrested in 1940 and died in prison in 1943. Lysenkoism held sway in the Soviet Union until the mid-1960s. Because of this, biological science in Russia and the other countries of the former Soviet Union still lag behind the West.

Muller got caught up in the controversy. He debated Lysenko in Moscow in December 1936 in front of an audience of about three thousand geneticists and collective farmers who were equally divided in their support for genetics or Lysenkoism. Muller accused Lysenko of practicing shamanism instead of science and called him a fraud. He was shouted down by the crowd. Realizing there was no future for genetics research in the Soviet Union, Muller looked for a safe way to exit the country.

He managed to escape by enlisting as a volunteer in the Spanish Civil War on the side of the Republican Army, which was supported by the Soviet Union. In this way, Muller thought that he could leave the country without endangering his Russian colleagues. Muller did physiological research on blood transfusions until the Republican Army was on the verge of defeat. Now Muller faced a dilemma. He couldn't go back to the Soviet Union, where he might be arrested or executed. He couldn't go back to Austin because he had received a notice from the university that he would have to stand trial in the faculty senate for violating a policy that required that editorials and columns in student publications be signed. Muller tried to find work at universities in Paris and Stockholm, but they had no openings. His friend from Rice, Julian Huxley, heard about Muller's situation, contacted the Institute for Animal Genetics at the University of Edinburgh, and arranged a position for Muller as a guest investigator.

Muller arrived in Edinburgh, along with about 250 strains of Drosophila, in September 1937. Among other research, Muller and a colleague looked at the relationship between radiation dose and mutation frequency and found that the same level of mutations is produced by a given dose whether the dose is administered over a month or half an hour. Muller applied the result to human beings, arguing that even small, diagnostic doses of radiation administered by doctors might be harmful. Physicians objected to Muller's conclusion, noting that the research was done on fruit flies not people and complaining that it undermined patient confidence. This was the beginning of a conflict with the medical community regarding radiation safety that would continue for the rest of Muller's life.

At Edinburgh, Muller met and began dating Dorothea "Thea" Kantorowicz, a refugee from Nazi Germany. The couple married in May 1939. Later that year, the Seventh International Congress on Genetics, an event planned by Muller, was held in Edinburgh. The congress had originally been scheduled to take place in Moscow but was moved due to Lysenkoism. For the congress, Muller wrote a paper titled "Geneticists' Manifesto" in response to the question "How could the world's population be improved most effectively genetically?" He also participated in a debate regarding the existence of the gene, for which there was still little direct evidence.

Shortly after the congress, war broke out in Europe. That put an end to basic scientific research in England and Muller was advised to return to the United States, which he did in 1940. Because of his communist sympathies, his job search was frustrating. The best he could do was a temporary research position at Amherst College in central Massachusetts. After the United States entered World War II, the term of the position was extended and his duties were expanded to include teaching. It was a frustrating time for Muller; he had no graduate students, no assistants, poor facilities and no financial support. He had a heavy teaching load that including teaching biology to uninterested military trainees. He told a colleague that he had tried to give his students a mnemonic to help them remember that red, oxygenated blood is on the left side of the heart. He told them that red is left, but the political meaning of the memory aid escaped his students.

His publication rate sharply declined, although he did manage to do a large-scale experiment on the relationship between aging and spontaneous mutations. Muller's main activities at Amherst were writing review articles and teaching undergraduates, an activity he didn't like. He served as a consultant on radiation genetics for the top-secret Manhattan Project, but the research could not be published and even Muller didn't know the purpose of his work. Nevertheless, Muller, living in a small upstairs apartment in an old frame house, enjoyed family life with Thea and a new baby daughter named Helen.

Muller's appointment ended in the spring of 1945, and he was not invited to join the faculty. He wrote letters to his friends asking if they knew of any position. He was desperate, and it showed. In one letter, he questioned whether he could continue living an academic life; it was so alarming that the recipient burned it. Fortunately, Indiana University came to the rescue. Fernandus Payne, a geneticist at IU who had worked in Morgan's fly lab, heard that Muller was available and sent biologist Tracy Sonneborn to meet with Muller to explore the possibility of him coming to IU. It is a great credit to Payne that he ignored the labels of communism, mental illness and being difficult and offered a job to Muller. Thus it was that IU landed one of the greatest scientists in the world.

In 1945, Muller and his family moved to Bloomington, where the fifty-four-year-old Muller would spend the rest of his life. IU's decision to hire Muller was quickly rewarded when he won the 1946 Nobel Prize for Physiology or Medicine. The Nobel citation reads, "For the discovery of the production of mutations by means of X-ray irradiation." The award dramatically elevated his status nationally and at the university.

The newly minted Nobel laureate started a graduate program at IU and taught three graduate-level classes per year. In collaboration with Brookhaven National Laboratory on Long Island, he initiated a variety of radiation genetics projects using neutrons and other particles. He considered problems in human genetics and introduced a new concept known as "genetic load." He and his students also studied spontaneous mutation rates under a variety of external environmental conditions such as oxygen- or nitrogen-rich atmospheres. He also developed methods of improving the detection of visible, sterile and lethal mutations.

One of Muller's students was the astronomer Carl Sagan, who spent the summer of 1952 working in Muller's lab. Sagan met Muller through a mutual acquaintance who showed the Nobel laureate a letter Sagan had written on the origin of life. Muller invited Sagan to come work for him that summer, and Sagan immediately accepted. The two hit it off and enjoyed a warm friendship, even attending a science fiction convention together. Sagan's wife, Ann Druyan, recalled that Muller treated Carl "with such kindness when he was the most naïve and gawky and dorky of young grad students."

Muller's time at IU was not without controversy. In 1948, he publicly attacked Lysenkoism in the Soviet Union at the International Genetics Congress in Stockholm. When he started enumerating the crimes against science committed by the Lysenko's followers, the delegates from the Eastern Bloc got up and walked out. The controversies multiplied during the Cold War and the McCarthy era. Some misconstrued Muller's warnings about radiation hazards in medicine, industry and nuclear weapons as attacks on national security. His past communist leanings led to suspicion, and he was called to testify before the House Un-American Activities Committee, testimony that remains secret. Muller recognized that he had been mistaken about the Soviet Union—that instead of a communist utopia, it was, under Stalin, a brutal dictatorship. Nevertheless, during the McCarthy era, he and his wife burned thousands of letters and other documents that might incriminate former students and colleagues who were sympathetic to the communist cause.

In 1957, Muller returned to the issue of eugenics. He argued that the past abuses of the American eugenics movement and the outright crimes of Nazi eugenics were not likely to be repeated. The old-style eugenics, based on dubious science and prejudice, was out. Muller advanced a new and more humane eugenics based on rigorous experimentation with the lofty goal of social equality. He proposed that prospective parents be granted the power to influence the genetic makeup of their children through the use of artificial insemination, donor eggs, adoption or other technologies. He hoped that people would learn to separate sexual activity from the genetic quality of their children just as they had learned, through the use of birth control, to separate sexual activity from reproduction. He envisioned a genetic enlightenment in which family planning would include a discussion of the desired genetic makeup of the child. Newspapers criticized Mullers views on eugenics, citing the Holocaust and the excesses of the American eugenics movement. Some accused him of trying to revive the eugenics movement of the 1920s that he himself had so roundly condemned. Near the end of his life, Muller decided that no eugenics was better than bad eugenics and declined to endorse a California sperm bank.

Muller retired from IU in 1964 and accepted a one-year appointment at the Institute for Advanced Learning in the Medical Sciences at the City of Hope, a cancer hospital in Duarte, California. Hermann Muller died of a heart condition on April 5, 1967, at age seventy-six, in Indianapolis.

HAROLD UREY

1893-1981

Field | Chemistry
Major Contribution | Won the Nobel Prize for Chemistry in 1934 for his discovery of deuterium, an isotope of hydrogen
Indiana Connection | Born in Walkerton, went to high school in Kendallville and attended Earlham College in Richmond

> *I looked for it* [deuterium] *because I thought it should exist. I didn't know it would have industrial applications or be the basis for the most powerful weapon ever known....I thought maybe my discovery might have the practical value of, say, neon in neon signs.*
> —*quoted in "Moon-Struck Scientist,"* New York Times,
> April 27, 1961

HAROLD CLAYTON UREY was born in the small town of Walkerton in the Amish region of northern Indiana on April 29, 1893. He grew up in rural poverty. According to Matthew Shindell's biography, *The Life and Science of Harold C. Urey*, when Urey was later asked what he remembered about his childhood, he gave a simple answer: "Poor. We were very poor. I remember terribly poverty stricken days."

Harold's father was Samuel Clayton Urey, an educator and a preacher, and his mother was Cora Rebecca Reinoehl. The family belonged to the German Baptist Brethren Church, also known as the Church of the Brethren, a religious sect similar to the Amish. Because of the group's practice of immersing the entire body during baptism, members were known as the "Dunkers." Daily life in the Urey family was organized around their faith. Mornings began with prayer and family worship. On Sundays, they rode a horse and buggy to the meetinghouse, where men and women sat apart from one another. They adhered to a strict dress code: men wore plain coats with no ties and women wore long skirts and long-sleeved blouses with high-necked collars.

Samuel had attended college and took a job in Walkerton as a school superintendent. The family soon had to move to Samuel's mother's farm in Corunna to help out with two sisters who had contracted tuberculosis. Unfortunately, Samuel got the disease himself and a doctor recommended a move to a drier climate. The family, which now included another son named Clarence, moved to Glendora in southern California, where Samuel worked at a packinghouse putting together wooden crates while Harold and Clarence played on the floor. On Sundays, he was an unpaid missionary minister.

The manual labor made Samuel's condition much worse. He realized he was dying and was ridden with guilt about leaving his wife, now pregnant with a third child, alone to raise the family. As a result, his mental health began to decline. After preaching a sermon in July 1897, Samuel snapped and, according to a newspaper report, tried to kill an old man. A judge committed him to an insane asylum, where doctors diagnosed him with tuberculosis meningitis; the disease had gone to his brain. Samuel spent eight months in the asylum while Cora supported the family by taking in laundry. While Samuel was away, Cora gave birth to a daughter named Martha.

About a year after his release, Samuel concluded that the California climate wasn't helping fight the disease and decided he'd rather die back home in Indiana. In June 1899, they returned to his mother's farm, where Samuel died six months later. For several years, Cora had to make it on her own. She was described by the family as "truly a strong, hardworking and brave pioneer mother and wife." Martha later poignantly wrote: "I have always thought I must have been the most unwanted child in the world. How could anyone want a baby at a time like that? Yet, as I grew up I felt loved and extra special. My dear wonderful mother was a remarkable woman."

Cora remarried in 1903 to a hired hand named Martin Alva Long, known simply as Alva. The family moved to Alva's forty-acre farm near Cedar Lake, Indiana, where they grew onions. Martha later said the farm "consisted of some low land down by a creek that emptied into the lake. There was some higher ground for pasture and other crops, a truck patch, and an orchard.

The house and barn were on top of a hill, and both were made of logs."
Harold recalled summer days spent weeding the onion field, swimming in
the lake and sleeping in the attic. Alva and the boys kept the family from
going hungry by catching fish from the creek. "It was a very pleasant life
on the whole," Harold later remembered. "Terribly hot in the summertime,
however, in the onion field."

Urey was educated at an Amish grade school, where he graduated at age
fourteen. Academically, he ranked next to last in his class of thirteen students
and barely passed the state's graduation exam. On graduation day, each
student gave a short speech about what they had to do to reach this point
in their education. Urey chose to speak about a topic he knew something
about: perseverance.

The next step was high school, but there was no school near the Cedar
Lake farm. The family used money from Samuel's life insurance policy to
pay for Harold's room and board with maternal relatives who lived near
Kendallville. Harold would spend the week in Kendallville and ride his bike
back to the onion farm on weekends.

Kendallville was a small town with a population of only five thousand, but
it gave Harold a chance to experience the larger world outside of the insular
Church of the Brethren. Although his relatives were also devout Christians
and members of the First Evangelical Church, they did not adhere to the
same rigid dress and behavior codes as the Brethren. Harold reinvented
himself in Kendallville, casting aside the dress, dialect and demeanor of the
Brethren. He abandoned his shyness, befriended girls and attended parties.
He started earning good grades and soon found himself at the top of his
class. In recognition, his classmates started calling him "Professor."

It was during his high school years that Harold began to turn away from
religion after reading the works of the agnostic freethinker Robert Ingersoll.
Ingersoll argued that the Bible was written not by God, but by men, and
therefore should not be taken literally. Harold eventually adopted much of
what Ingersoll said, defining himself as an atheist because he did not believe
in "a private God that listens to the prayers of anyone." Urey was inspired
by nature, saying: "To me, the enormous universe and all the things in it
are the source of my wonder, and I need no God to increase this wonder
at all." However, Urey did believe that Christianity, especially the Ten
Commandments and the teachings of Jesus, had helped civilize the West
and pave the way for science.

After graduating from high school in 1910, Harold decided to pursue
teaching as a career. He went to Earlham College, a Quaker school

in Richmond, Indiana, where he earned his teaching certification in 1911 and then spent a year or so teaching in small rural schoolhouses. Meanwhile, Alva had moved the family to an eighty-acre homestead near Big Timber, Montana. He bought the land at a cheap price through the Carey Act, which provided irrigated land to farmers. Unfortunately, the soil was poor and the winters were brutal. Less than four years later, they sold everything and left.

Harold rejoined the family in Montana, where he continued teaching for two years at one-room schoolhouses around Yellowstone National Park. To save money, Harold lived and ate with the families of his students in return for tutoring. While teaching in a mining camp, the son of the family he was living with left for college. Harold thought to himself: "If perhaps I were going to get ahead in the world, I should go to college also." In the fall of 1914, as war was erupting in Europe, Harold enrolled at the University of Montana in Missoula.

His first choice for a major was psychology, but the department had a policy of not admitting freshmen, so he enrolled instead in biology and chemistry classes. There, he came under the influence of several notable professors, the most important of which was Archie Wilmotte Leslie Bray, a Cambridge-educated biologist. Looking back, Urey said of Bray: "I soon realized that under him I was getting a Cambridge education and was gradually changing from a little country boy to more nearly a man of the world....Professor Bray was just a splendid, model teacher who opened up the whole fascinating world of science to me." Under Bray's tutelage, Urey majored in biology and chemistry, graduating from the University of Montana in 1917 just as the United States officially entered World War I.

Urey joined the war effort as part of the chemical industry and went to work in Philadelphia, where he helped prepare the chemicals used in ammunition, including the toluene for trinitrotoluene, or TNT. It was simple industrial chemistry but vital to the war effort. According to Urey, this is when he stopped being a biologist and started being a chemist.

After the war ended in 1919, Urey returned to the University of Montana as an instructor in the chemistry department. He eventually realized that if he wanted to climb the academic career ladder, he would need to earn a doctorate. The chair of the chemistry department wrote Urey a glowing letter of recommendation that resulted in Urey being offered a graduate fellowship at Berkeley. In 1921, at age twenty-eight, Urey entered the doctoral program in chemistry at the University of California at Berkeley, a school that was quickly becoming one of the top research universities in the country, especially in the area of physical chemistry. Urey later called his two years as a graduate student among the most inspiring of his life.

Urey became an expert at using a spectroscope, an instrument that he would later use to make his greatest discovery, and earned his doctorate in 1923. Realizing he still had much to learn about the structure of atoms and molecules as revealed by the new field of quantum mechanics, he accepted a postdoctoral fellowship to work at physicist Neils Bohr's Institute of Theoretical Physics in Copenhagen, the world's leading center for quantum physics. At Copenhagen, Urey worked alongside many of the leading physicists and chemists of the day, but he quickly realized that he did not have the necessary mathematical talent to be a theoretician. In an unpublished autobiography, Urey confessed, "I tried to do some theoretical physics at Bohr's Institute....Perhaps the most important thing that I learned was that these friends of mine at the Institute…would be able to do a much more thorough job in theoretical physics than I could possibly do, and that my strength in science would lie elsewhere."

Upon returning to the United States, Urey accepted a position at Johns Hopkins University in Baltimore. Although Urey did no original research in quantum theory, he did have a knack for teaching about it and started a new course on atomic physics. To reach a wider audience, in 1929 Urey published

The Niels Bohr Institute in Copenhagen, Denmark. *Wikimedia Commons.*

a book with Arthur Ruark titled *Atoms, Molecules, and Quanta*, one of the first English-language texts that applied the new quantum mechanics to atomic and molecular systems. The book helped establish Urey as one of America's leading experts in the field of quantum mechanics. In the same year Urey was named editor of the new *Journal of Chemical Physics*, a journal whose intent was to publish papers about research spanning the two disciplines.

Urey's four years at Johns Hopkins were, by his own assessment, not very productive. He managed to publish only one or two papers per year and later refused to cite any of them, saying that he would not reference papers that were incorrect. The department was weak in Urey's areas of interest, and the research facilities did not fit his needs. Urey had a particular dislike for Joseph Ames, a university provost and head of the physics laboratory. (NASA's Ames Research Laboratory is named after him.) Urey considered Ames to be old-fashioned, imperious and uninterested in the new quantum physics. Ames became president of the university in 1929, and according to Urey, the university began to decline. Urey even wrote to a colleague claiming that "this man destroyed the university." So when New York's Columbia University offered Urey a position as an associate professor of chemistry, he jumped at the opportunity.

Although Urey's professional life languished during the Johns Hopkins years, he had better luck in his personal life. In the summer of 1925, Urey traveled to Seattle to visit his mother. On the way, he stopped in Everett, Washington, to say hello to a friend named Kate Daum, a woman he had known as a student at Montana. Kate, in turn, introduced Urey to her sister Frieda. Urey and Frieda hit it off. He liked her because she was a fellow midwesterner, from Kansas, and as a bacteriologist, she understood science. She was impressed by his intellect and his time in Europe. The couple spent the next two weeks hiking in the Cascades. He wanted to get married immediately; she was more cautious—after all, they had just met. The courtship continued by mail for a year until they were married at her family home in Lawrence, Kansas, in June 1926. The couple went on to have four children: Gertrude Bessie in 1927, Frieda Rebecca in 1929, Mary Alice in 1934 and John Clayton in 1939. By all accounts, it was a happy marriage.

At Columbia University, Urey had more money, better equipment and a research assistant. In particular, there was a spectroscope nobody was using in the basement of Pupin Hall. Urey and his assistant, George Murphy, laid claim to the instrument and modified it to their own design. Urey's work focused on isotopes. We now know that isotopes are different forms of the same chemical element—the difference being the number of neutrons in the nucleus. But the neutron wasn't discovered until 1932. In 1931, scientists thought that the nucleus of an atom consisted of protons and nuclear electrons. Isotopes differed in mass because they had different numbers of nuclear electrons.

There was only one known form of hydrogen at the time—hydrogen with a single proton in its nucleus. But Urey suspected there was a heavier form of hydrogen. He had constructed a chart of isotopes showing the number of nuclear electrons plotted against the number of protons. The known nuclei of the light elements made straight line segments, but there was an empty spot for hydrogen, suggesting a missing isotope. Urey's suspicion was bolstered by other research. But if the heavier isotope did exist, it would be extremely rare and difficult to identify. Undaunted, Urey set out to find it.

On Thanksgiving Day 1931, the spectral signature of the heavy isotope of hydrogen was confirmed. Urey arrived home that evening several hours late for Thanksgiving dinner. As he entered his apartment on Claremont Avenue, he turned to his wife and said, "Frieda, we have arrived!" Unlike isotopes of a given element whose properties were usually similar, heavy hydrogen's chemical and physical behavior was much different from that of regular hydrogen. This prompted Urey to give the isotope an actual name: deuterium. The discovery prompted a frenzy of research; between 1931 and 1934, more than two hundred scientific papers were published about deuterium. It affected research in physics, chemistry, biology and medicine and helped usher in the atomic age.

For his discovery of "heavy hydrogen," Urey was awarded the Nobel Prize for Chemistry in 1934, but he missed the Nobel Prize ceremony that December so that he could be present at the birth of his daughter Mary Alice. Although Urey was given the entire monetary award, he shared the prize money with two colleagues who had helped with the experiment. Urey used most of the remaining prize money to build a new house in Leonia, New Jersey, just across the river from Manhattan.

Sharing money with people wasn't Urey's only generous act. As early as 1938, he began helping Jewish scientists escape Europe and finding them positions in the United States. When Mussolini introduced anti-Semitic laws in Italy, Urey helped physicist Enrico Fermi; his Jewish wife, Laura; and their two children come to America and convinced them to settle in Leonia just down the street from the Ureys.

During the war years, Urey played an important role in the Manhattan Project, the effort by the United States to build an atomic bomb. The history of the Manhattan Project is described in the Pulitzer Prize–winning book

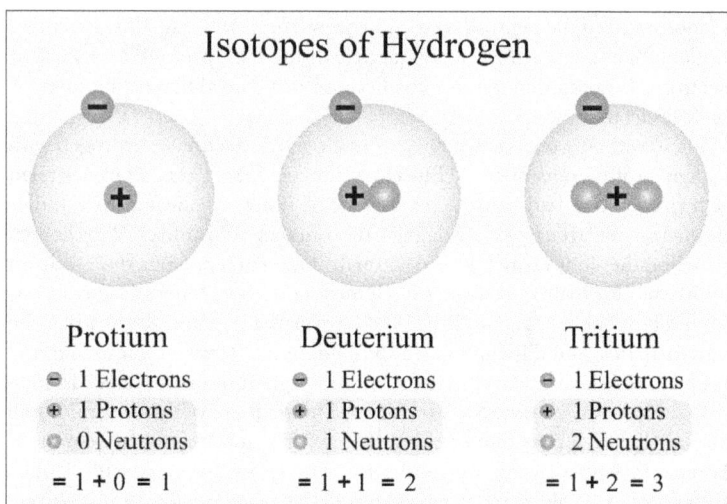

Isotopes of Hydrogen

Protium	Deuterium	Tritium
⊖ 1 Electrons	⊖ 1 Electrons	⊖ 1 Electrons
⊕ 1 Protons	⊕ 1 Protons	⊕ 1 Protons
○ 0 Neutrons	○ 1 Neutrons	○ 2 Neutrons
= 1 + 0 = 1	= 1 + 1 = 2	= 1 + 2 = 3

The three isotopes of hydrogen with differing numbers of neutrons in the nucleus. Urey discovered deuterium. *Shutterstock.*

The Making of the Atomic Bomb by Richard Rhodes. At the time, it was known that a rare isotope of the element uranium would undergo nuclear fission when hit by a neutron. (This is the so-called splitting of the atom.) The problem was that this rare isotope had to be separated from the much more plentiful isotopes of uranium, isotopes that would *not* undergo fission. The separation of isotopes of a chemical element is an extremely difficult task because the isotopes are chemically identical. Thus, isotopes cannot be separated by means of a chemical reaction. The only distinction between isotopes is a small difference in mass.

Harold Urey was one of the world's leading experts on isotopic separation and began work on the uranium isotope problem in May 1940. Several different methods of separation were investigated, including a centrifuge technique, thermal diffusion and gaseous diffusion. A fourth method by electromagnetic means was directed by physicist Ernest Lawrence in California. By May 1943, Urey was responsible for all the research on uranium separation (except the electromagnetic method) and on heavy water. By the end of the year, Urey had hundreds of people working for him.

But Urey hated being an administrator and was exhausted by the diffusion work. And he did not get along with his bosses, namely General Leslie

Groves, head of the Manhattan Project. Groves told a story about having lunch with the chemist and noticing that Urey was unable to hold a glass of water without it shaking. In an interview, Urey admitted, "I was most unhappy during the war. I had bosses in Washington who didn't like me, and I had people working for me who didn't like me. Imagine a more miserable situation—where you can't resign, but nobody wants you around!" Urey then went on to say that he was "very close to a nervous breakdown." In February 1945, he left the Manhattan Project.

Urey was also unhappy at Columbia, complaining that there existed "a great deal of personal jealousy between people in the department instead of friendly boosting of each other. At Columbia I had very few friends. They weren't really friendly to each other." And then there was the political divide—Urey was a liberal Democrat while the other members of the department were conservative Republicans. As a student from the time recalled, "In 1936 when Roosevelt was running against Landon, Urey sported a Roosevelt button. But he was the only member of the chemistry department who did. The others all had Landon buttons." Finally, Urey found the department resisted students who didn't fit the mold of a white Protestant male. Jewish students struggled for acceptance, and women were treated as second-class citizens. All this prompted Urey to look for a new home.

In 1945, Urey and some colleagues negotiated a deal for a new Institute for Nuclear Studies to be established at the University of Chicago. Urey moved to Chicago and bought a home in the Hyde Park neighborhood near the university. His used his platform as a Nobel laureate to promote his political beliefs. Urey's liberal politics led him to speak out against war and in favor of world government. He publicly supported Julius and Ethel Rosenberg, although he suspected they were guilty of spying. During the McCarthy era, he was called before the House Un-American Activities Committee.

Shortly after moving to Chicago, Urey read a book titled *The Face of the Moon* by Ralph Baldwin. He fell in love with the moon and, more broadly, the solar system—a love that would last for the rest of his career. He studied the chemical composition of meteorites, developed theories about the abundance of chemical elements on earth and created an entire new field he dubbed cosmochemistry. He summarized his work in a 1952 book titled *The Planets: Their Origin and Development.*

Urey was also interested in research about the origin of life on Earth—work that would involve him in one of the most amazing experiments

Harold Urey working in his lab. *U.S. Department of Energy/Wikimedia Commons.*

ever performed. In a paper, he speculated about the composition of the atmosphere on the early Earth and concluded that it was rich in hydrogen, methane, ammonia and water. He further suggested that if a spark of energy were introduced into the mixture, organic compounds might result and proposed an experiment to test the idea. Urey gave a seminar presentation on the subject and, afterward, went back to his office, where he was visited by a graduate student named Stanley Miller. Miller said that he wanted to do his doctoral dissertation on the subject. Urey thought that "it might not turn out well for a dissertation, but Stanley could not be discouraged."

With Urey's help, Miller designed an experiment to simulate the conditions Urey had proposed. The experiment consisted of a large glass flask into which methane, ammonia, hydrogen and water were added. Energy in the form of an electrical spark was introduced, the water was heated and the experiment was allowed to run. After a couple of days, the mixture had turned a yellowish brown. Analysis of the resulting goo showed the presence of several amino acids and other building blocks of life—an almost unbelievable result.

At the crowded seminar where Miller presented his results, Urey sat in the front row. During the question-and-answer session at the end, Enrico Fermi looked at Urey and asked: "I understand that you and Miller have demonstrated that this is one path by which life might have originated. Harold, do you think it was *the* way?" Urey answered: "Let me put it this way, Enrico. If God didn't do it this way, he overlooked a good bet!"

In May 1953, a paper describing the experiment appeared in the journal *Science*. Miller was listed as the sole author, although a footnote acknowledged Urey's assistance. This established Miller as a leader in origin of life research and helped him avoid being drafted into the military during the Korean War. As it turns out, Urey may have been wrong about the composition of the primordial atmosphere—the question is currently under debate in the scientific community. But the experiment has been repeated using different starting chemicals, and the result is always basically the same.

Personal conflicts prompted Urey's departure from the University of Chicago, and in November 1958, he moved to La Jolla, California, to join the newly formed University of California–San Diego (UCSD). Urey helped build the science faculty at UCSD and continued his research, publishing over one hundred scientific papers. When someone asked him why he continued to work so hard, he jokingly replied, "Well, you know I'm not on tenure anymore."

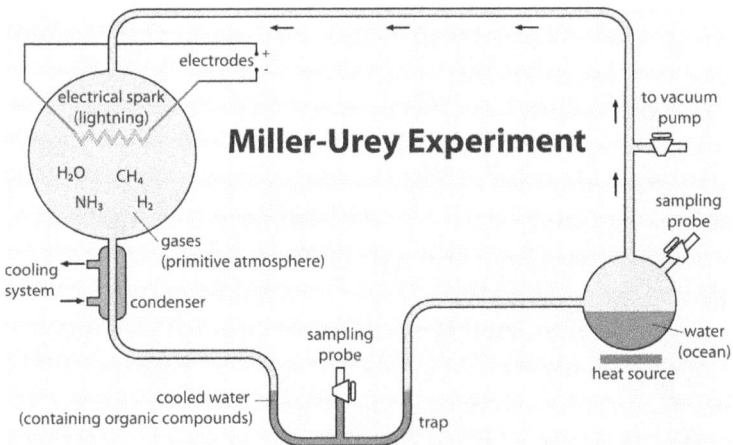

Diagram of the Miller-Urey experiment where Earth's primordial chemicals were mixed together and subjected to electrical sparks, resulting in the production of organic molecules. *Shutterstock.*

Urey continued to work into his eighties, publishing his final two papers in 1977 at age eighty-four. In an annual birthday letter written that year, Urey laments, "I am feeling my old age every day. My hands quiver, I wobble when I walk, my eyesight is bad and I can't remember things so that I cannot keep up with the literature, hence I can do no scientific work at all." Parkinson's disease caused the quivering, and macular degeneration damaged his eyesight. Two years later, Urey wrote to an artist who had painted his portrait and said, "Now I take joy in sitting on my beautiful patio and viewing my flower/orchid garden."

Harold Urey, one of the greatest chemists of the twentieth century, died on January 5, 1981, at age eighty-seven of an apparent heart attack. His body was cremated in La Jolla and then buried back home in Indiana at the Fairfield Cemetery in DeKalb County, where Frieda joined him eleven years later.

6

ALFRED KINSEY

1894-1956

Field | Biology
Major Contribution | Pioneer in the field of human sexuality
Indiana Connection | Professor at Indiana University from 1920 until
  his death in 1956

> We are the recorders and reporters of facts—not the
> judges of the behaviors we describe.
> —*Alfred Kinsey*

ALFRED CHARLES KINSEY was born on June 23, 1894, in Hoboken, New Jersey, across the river from New York City. His father, Alfred Seguine Kinsey, was a professor at the Stevens Institute of Technology. His mother, Sarah Ann Charles, had only a fourth grade education but was a sweet and gentle woman. A younger brother, Robert, was born in 1908. The family was dominated by Alfred Seguine, who was a stubborn, domineering and egotistical husband and father.

As outlined in Jonathon Gathorne-Hardy's biography *Kinsey: Sex, the Measure of All Things*, Kinsey's childhood was unhappy for three reasons: disease, poverty and religion. The first ten years of Kinsey's life were dominated by disease. In addition to the usual childhood illnesses of measles and chicken pox, Kinsey contracted rickets, rheumatic fever and typhoid. Rickets is caused by a poor diet lacking in Vitamin D, resulting in the softening and weakening of the bones. It left Kinsey with a double curvature of the spine. Rheumatic fever is an inflammatory disease that develops when strep throat or scarlet fever isn't properly treated. It is a painful and debilitating disease that leaves its victim bedridden, feverish, aching and exhausted. In 1903, Kinsey contracted typhoid, a bacterial infection often caused by contaminated food and water. At the time, there was no treatment, and the mortality rate for young children was close to 60 percent. But young Kinsey recovered. As a positive side effect of the typhoid, he was cured of the rheumatic fever, probably because the high temperatures of 104–6°F killed off the streptococcal bacterium.

A second prevalent feature of Kinsey's childhood was poverty. His father gave his mother a small and inadequate allowance each week with which she would have to make do. She often sent her son to the store to beg for more credit, an act that he found humiliating. Kinsey never forgot his childhood poverty. As an adult, he was extremely frugal: he wore mended clothes, always tried to buy on the cheap, never allowed his wife to run up a charge account and never accumulated any debt.

The final pervasive feature in Kinsey's childhood was religion. The family belonged to a strict Methodist sect, and Kinsey's father was one of its most extreme members. Sundays were the worst. The family went to Sunday school and church, where young Alfred had to endure three long services. Absolutely nothing else was allowed: no cooking, newspapers or entertainment. The only permissible activity was prayer. And father Kinsey was not merely content in enforcing his view of morality on his own family; he zealously tried to impose it on the entire community. Alfred Seguine would deliberately send his young son out to try to illegally buy cigarettes. If young Alfred was successful, Alfred Seguine would inform the police about the shopkeeper.

The young Kinsey found a refuge from the dreariness in music. The family occasionally visited relatives who had one of the new-fangled Edison talking machines that recorded sound on rotating cylinders. The magic cylinders also played music, and Kinsey was entranced. Noticing the reaction, his parents provided piano lessons beginning at age five. By age sixteen, he was an accomplished pianist, giving recitals at home, playing Beethoven's *Moonlight Sonata* at a school concert and performing a piece by Chopin at his high school commencement. Kinsey's love of music continued throughout his life.

At age fifteen, Kinsey started high school. His most influential teacher was Natalie Roeth, a young and enthusiastic biology teacher, who introduced Kinsey to Darwin's theory of evolution, a controversial topic she taught unflinchingly. Evolution became the dominant idea in Kinsey's intellectual life and caused him to begin questioning his own religious beliefs. Kinsey always credited Roeth as the reason he became a scientist.

In high school, Kinsey also discovered he was smart. If he couldn't compete on the playing field (he found sports boring), he could certainly compete and win in the classroom. He immersed himself in his schoolwork, often working late into the night, and developed an extraordinary ability to concentrate. A classmate later recalled that Kinsey had a nickname: the Great Scientist. It was an appropriate label, since Kinsey wrote his first scientific paper at age sixteen. The paper, innocently titled "What Do Birds Do When It Rains?," was based on hours of careful observations. To his and Roeth's delight, it was published in a nature journal.

By the time Kinsey graduated from high school in 1912, he had decided to become a scientist and study biology. But his father had instead decided his oldest son would attend the Stevens Institute, where the tuition was free, and become an engineer. The father won, and Kinsey spent the next two years at the Stevens Institute—years that he would later characterize as wasted and frustrating. The reluctant student nevertheless worked hard and did well in nearly all his subjects except for advanced math and mechanics, which he had to retake. His only solace at Stevens was playing the piano in the orchestra.

By the end of his second year, Kinsey had had enough. He confronted his father, telling him he was leaving the Stevens Institute and going to Bowdoin College in Brunswick, Maine. The argument dragged on for weeks, but Alfred Seguine ultimately gave in. Kinsey's tuition would be partially paid by an annual $200 scholarship, and he would work to pay the balance. In September 1914, Kinsey donned a new suit and left for Bowdoin. It was essentially the end of Kinsey's relationship with his father, who rarely saw or helped his oldest son again.

Because of his two years at Stevens, Kinsey entered Bowdoin as a junior and once again threw himself into his work, spending nearly every waking moment in intense study. When he needed a break, he played the piano. He didn't attend dances, parties or football games and had no social life or real friends. And yet he seemed happy and well adjusted. He was also quite handsome, with curly blond hair, blue eyes and a pleasant smile. Women noticed him, but he was too shy and busy to reciprocate.

Kinsey graduated magna cum laude from Bowdoin and won a scholarship to Harvard University. There, the most important figure in Kinsey's academic life was his supervisor, William Morton Wheeler, an entomologist and the most famous biologist of his day. Wheeler smoke, drank and was an atheist. His influence probably contributed to Kinsey's gradually waning religiosity. Wheeler's lectures on insects fascinated Kinsey, who was searching for a dissertation topic. In the spring of 1917, he discovered the gall wasp—the little bug that would occupy him for the next twenty years.

The gall wasp is a small insect a few millimeters long, about the size of an ant. Although it's called a wasp, it can neither fly nor sting. Usually, the eggs are laid in oak trees and deposited with a poison that causes a gall, an abnormal tissue outgrowth on a tree similar to a tumor in animals. The eggs sit protected in the gall, sometimes for months, until they hatch into larvae. The larvae cause the tree to form a new gall, where the larvae eat and live for a time ranging from a month to three years. Finally, the pupae emerge, grow into adults, mate, lay eggs and die—all within a few days or weeks.

Over the next two years, Kinsey collected thousands of galls and gall wasp specimens, mostly from the Boston area, and took numerous measurements of each one. In the spring of 1919, he finally turned in his dissertation to Wheeler. The professor was astonished and told a colleague that to his great surprise, this graduate student whom he barely knew was of a very high caliber. In June 1919, Kinsey graduated from Harvard with a doctorate in biology.

Kinsey's outstanding work won him a grant that paid travel expenses for as long as he could make the money last. He set out on a ten-month solo journey across America, by train and on foot, to collect gall wasps. He visited

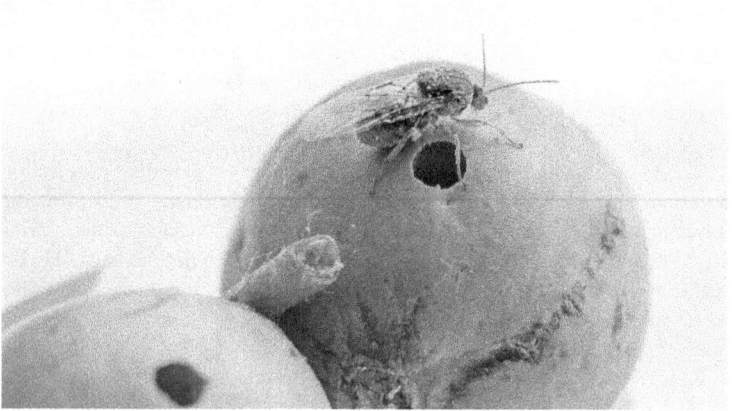

Parasitic gall wasp emerging from its home. *Shutterstock.*

thirty-six states, spending most of his time in the mountains of California, Arizona and Texas and traveling a total of 18,000 miles, 2,500 of which were on foot. By journey's end, he had collected 300,000 specimens.

During his long expedition, Kinsey was, with Wheeler's help, also searching for a job. In April 1920, Wheeler was contacted by Carl Eigenmann, head of the zoology department at Indiana University, who asked if he could recommend someone for a position. Wheeler told Eigenmann about Kinsey, and a job offer followed. Kinsey had been in Indiana in August 1918 serving as a woodcraft instructor at the Culver Military Academy in northern Indiana. He didn't like it—the state was too hot, too humid and too flat. But the glaciers that had smoothed the topography of the northern part of the state didn't quite make it as far south as Bloomington. Kinsey found that out when he took a few days off from camp work to visit. He liked the rolling and wooded countryside and loved the generous $2,000 salary (about $26,000 in today's dollars) he was offered. He accepted the job. Eigenmann was delighted and wrote to the university president saying that Kinsey's Boy Scout and YMCA work was evidence that he would be a safe choice for an instructor. That part didn't quite turn out the way Eigenmann expected.

And so it was that Alfred Kinsey, age twenty-six, wound up at Indiana University. In 1920, IU was not the university it is today. It was small, with only 2,356 undergraduates, nearly all from Indiana. The faculty was not particularly distinguished, and only a few departments offered doctoral degrees. The bucolic setting was evidenced by the presence of cows that wandered around the campus eating grass and producing milk for the students.

Kinsey met his future wife that autumn at a zoology department picnic at Spring Mill State Park, not far from Bloomington. He had actually met her, although very briefly, on three other occasions, but didn't notice her. But she had noticed him and later admitted that she was instantly attracted to him. Her name was Clara Bracken McMillen. She was twenty-one, short (five feet, two inches), black-haired and athletic. And Clara was smart—she had majored in chemistry, was a Phi Beta Kappa and had been elected to Sigma Xi, the scientific honorary. At the picnic, she got Kinsey's attention by suddenly leaving the zigzag path and hopping directly down the hillside. As he lit a fire to prepare some food, she joined him.

Although Kinsey hated sports, he took her to a basketball game that winter, where they got engaged—and never went to another game. They were married on June 21, 1921, in Brookville, Indiana, where her grandparents lived. Kinsey called her "Mac," and the marriage was a good match. She was his intellectual equal, was frugal with money and tolerated

Gate at Indiana University in Bloomington, Indiana. *Shutterstock.*

his eccentricities. She also understood that his work came first. They shared many interests and, most importantly, loved each other. Kinsey later told an associate that he had been in love three times in his life and that the first time had been with Mac.

The couple would go on to have four children: Donald in 1922, Anne in 1924, Joan in 1925 and Bruce in 1928. Kinsey was an involved father; he changed diapers, gave baths and fed bottles. But in June 1925, Donald, the oldest child, got sick. His eyeballs were protruding, a sign of an overactive thyroid. They took him to the Mayo Clinic, where he was operated on in September. It turned out that Donald also had diabetes, a rare disease in children, that was not diagnosed until March 1926. By that time, the little boy was very sick and fell into a coma. He died early in April at age four. Kinsey was devastated by the loss and cried uncontrollably. Mac's sorrow was even deeper.

In the 1920s, Kinsey wrote a successful and financially lucrative biology textbook, but his main focus continued to be research on the gall wasp. He traveled all over North America collecting thousands of galls and wasps. Sometimes the expeditions lasted for several months. When he returned, the specimens had to be prepared. This involved gluing the wasp to the tip of a stiff piece of paper, impaling the paper on a two-inch steel pin and attaching a tiny label. The pinned wasps were then mounted in boxes—eight hundred to a box, all facing to the right—and the boxes were carefully stored. This tedious work was done by young women working their way through college, earning about thirty cents an hour. Kinsey worked for long hours carefully examining every specimen under a dissecting microscope and recording twenty-eight different measurements.

Kinsey's work on gall wasps culminated in two scholarly books on the subject. It was recognized as the work of an outstanding scientist, and he was invited to speak at meetings of the American Association for the Advancement of Science. In October 1937, he was made a "starred scientist," meaning that the leading biologists of the country had voted and named him a top scientist in his field. His name would have a star placed beside it in *American Men of Science*, a book that gives biographical profiles of the country's leading scientists. At the time, IU had only four starred scientists. Kinsey became the fifth, resulting in an elevation of his status on campus.

Not many people were interested in gall wasps—only a dozen or so biologists ever examined his books. His gall wasp books sit today largely untouched deep in the stacks of the IU library. He later confided to an associate that although "studying gall wasps might have made some important contributions, it would never have caused people to beat a path to his door." Kinsey was right. Had he continued with his gall wasp research, his name would be unknown outside the small entomology and evolutionary biology community. But Kinsey was about to turn his attention to a subject that virtually everyone in the world was interested in: sex. One would think that the topic of sex was better suited for a campus like Berkeley rather than Bloomington, but this sleepy little town on the edge of the Bible Belt was about to become the nation's hotbed for sexual research.

By 1936, Kinsey had grown tired of IU. He complained about the university's leadership and told an associate that "the whole university is in a mess....I would leave at the first opportunity offering comparable recompense and research opportunities." Then, as 1937 rolled over into 1938, two events occurred that changed everything. First, the Indiana state legislature passed a mandatory retirement law and seventy-six-year-old longtime IU president William Lowe Bryan was out. In January 1938, thirty-five-year-old Herman Wells was appointed as the new president of IU. He was rotund, jovial, unapologetically progressive and highly intelligent. During his long tenure, Wells transformed IU from a mediocre university into a major research institution. He was a champion of academic freedom and courageously stood by Kinsey during the coming controversies.

The second event took place in November 1937 when a film titled *Forbidden* was shown on campus. It was a sex education film and students flocked to see it. It caused an uproar. The school newspaper, the *Daily Student*, ran an editorial demanding a sex education class at IU. But the paper didn't call it a "sex education" class, it called the prospective course a "marriage class." It is not clear whether the students approached Kinsey about the class or Kinsey approached the students—there are conflicting accounts of exactly what transpired. Nevertheless, fourteen student leaders presented a petition in support of the course to President Wells on May 14, 1938. Wells put Kinsey in charge of the class.

Marriage courses were not new. By 1938, about 250 universities offered them. The IU course would be open only to seniors or graduate students, married or engaged students and faculty. Both sexes would attend the class together. The class would consist of six biology lectures by Kinsey and six lectures, by other faculty members, on different aspects of marriage, including economics, sociology, psychology, legalities and religion. In June, Wells presented the proposed class to the trustees. They were wary—one trustee asked to be counted absent—but they agreed, asking only that no publicity be given to the class.

The marriage class was first offered in the summer of 1938. To understand its significance requires some historical context. At the time, public knowledge regarding human sexuality was abysmal. It was not uncommon to find coeds who didn't know where babies came from. Some thought that kissing could make you pregnant. The boys were equally ignorant. A 1939 –40 survey of high school boys ages twelve to eighteen found that 96 percent didn't know what the word *masturbation* meant (40 percent thought the act caused insanity), 91 percent didn't know what the word *virgin* meant and 27 percent didn't know what the word *intercourse* meant. The young men also displayed an alarming lack of knowledge of female physiology. Only 29 percent knew about menstruation or that the egg came from the mother (3 percent thought the hospital provided the egg).

Kinsey laid the blame for this profound state of ignorance squarely on religion. By this time, Kinsey's own religious faith had been washed away by science. He deeply resented the repressive power religion exerted over human sexuality and raged against it. "The whole army of religion is our central enemy," he told a friend. Kinsey refused to have anyone who was actively religious on his staff.

The first marriage class at IU had an enrollment of ninety-eight students; twenty-eight men and seventy women. Kinsey began the class with a rhetorical question: "Why offer a marriage class?" Then gave an answer: "Society has been responsible for interfering with what would have been normal biological development…leading to a scandalous delay of sexual activity which led to sexual difficulties in marriage…ignorance of copulatory techniques…ignorance of satisfactory contraceptive devices…concepts of sex all wrong." Kinsey's lectures discussed, in a frank and direct way, sex in the animal kingdom, human reproductive anatomy, masturbation and techniques of foreplay and intercourse. As one might imagine, the students were transfixed. One listener later claimed that you could hear a pin drop. By the fall of the year, enrollment in the class had doubled.

Kinsey was not known for his sense of humor, but every now and then, he came out with a zinger. Once, in an evolution class, he asked a coed what part of the human body could enlarge a hundred times. Aware of Kinsey's sex research, she answered by saying that she didn't know, adding, "You have no right to have asked me such a question in a mixed class." Kinsey stared at her and responded: "I was referring to the pupil of the eye, and I think I should tell you, young lady, that you are in for a terrible disappointment."

Kinsey solicited written feedback from his students and used it to improve the class. He also offered to have individual conferences with students regarding their sexual problems. But at these conferences, he also began doing something that was truly revolutionary—he began asking students for their sexual history. Soon, he was doing this openly. At the end of each lecture, he asked the students to volunteer their sexual history. And they did. This was uncharted territory—no one had ever done research on what people actually did sexually.

Over time, the content of the class changed, focusing less on how a healthy sex life enriches marriage and more on the broad range of human sexual experience. As the content evolved, the controversy increased. Criticism of the marriage course reached a critical mass in the summer of 1940 when Bloomington's Ministerial Association got involved and complained to the IU administration. Instead of removing Kinsey from the marriage course, Wells offered him a choice: Kinsey could teach the course or take the histories, but not both. Kinsey chose the sex histories. He taught the marriage course for the seventh and last time in the summer of 1940.

Relieved of the burden of the marriage course, Kinsey threw himself into the work of collecting sexual histories, approaching the task with the same unbounded enthusiasm he previously displayed for collecting gall wasps. He worked twelve-to-sixteen-hour days, six or seven days a week. He obtained financial support from the National Research Council and the Rockefeller

Foundation and built a team consisting of psychologist Wardell Pomeroy, anthropologist Paul Gebhard and statistician Clyde Martin.

To collect the histories, Kinsey and his team interviewed subjects and used a questionnaire consisting of 350 items. Beginning with socioeconomic data, it took Kinsey about twenty minutes to get to the first sex question. As an example, Kinsey asked how many orgasms the subject had in an average week and how were those orgasms achieved. The team was trained to be completely nonjudgmental about the answers.

As he had done for the gall wasp, Kinsey developed an elaborate coding system that condensed twenty-five pages of questions down to a single sheet of paper. Each sheet of paper was divided into 287 little rectangular boxes, each containing a coded answer to a question. The symbols meant different things depending on which box, or part of the box, it was recorded in. The code was not written down until many years later. Only Kinsey and his team of interviewers knew the meaning of the symbols. The sheet was totally incomprehensible to anyone who looked at it, thus preserving the confidentiality that Kinsey promised.

How did Kinsey guard against people lying about or exaggerating their sex life? Nothing is foolproof, but he found that maintaining constant eye contact and asking the questions rapidly made lying more difficult. There were also cross-checks built in to the questionnaire, asking a question in several different ways for example, that revealed inconsistencies. Finally, human sex lives tend to follow a pattern; a broken pattern sent up a red flag. Still, people could lie, and Kinsey admitted as much, saying that his results should be viewed as an approximation to the truth.

Kinsey set a rather unrealistic goal of collecting 100,000 sexual histories. He and his team fanned out across the country taking histories wherever they could. In 1947, the sex research group was organized legally into a separate and independent body from the university. This had been first suggested by Wells, who, although he supported the research, wanted to put some distance between IU and Kinsey's work. They all knew that controversy was coming. Kinsey liked the idea because it meant that he would be free from the inspection and oversight that a state university is subject to. Thus it was that Kinsey's team became known as the Institute for Sex Research.

By that time, Kinsey had enough histories to begin organizing and analyzing the data and publishing the results. The first book, a scientific treatise, took about two years to write. Wells made two suggestions regarding publication: they choose an established medical publisher to ensure the scientific integrity of the work and they publish when the state legislature

Staff of the Institute for Sex Research at Indiana University in 1953. Kinsey is seated in the center. Paul Gebhard is seated to his left. *Smithsonian Institution Archives/Wikimedia Commons.*

was not in session to soften any political blowback. As the publication date approached, things at the institute became more frantic and the press poured into Bloomington.

The book, titled *Sexual Behavior in the Human Male*, came out in early 1948. It was a sensation, quickly rising to the top of the *New York Times* bestseller list. The 735-page book was based on over five thousand male sexual histories. One topic that attracted much attention was homosexuality, with the book citing three notable statistics: 37 percent of the men had at least one homosexual experience, 10 percent were more or less exclusively homosexual for three or more years and 4 percent were exclusively homosexual for their entire lives. To measure this, Kinsey had developed a Likert-type, seven point, 0–6 scale. The zero indicated that the subject was exclusively heterosexual and the six meant exclusively homosexual, with the other numbers representing a mix of heterosexual and homosexual activity.

The reaction from the popular press was generally positive. The *New York Times* thought the book, although shocking to some, would have a healthy effect. The *Washington Post* argued that the book would, over the long term, lead to changes in sex laws and social attitudes. A Gallup poll found that of those who had heard about the book, 78 percent thought it was a good thing while 10 percent disapproved. But there were criticisms. No African Americans and few elderly people were included in the sample. Kinsey said simply that the sample from those groups was inadequate. Another criticism was that the sample was made up of volunteers and so it was not scientifically representative of the population. Kinsey countered by arguing that the large sample size made it representative. Subsequent calculations that adjusted the sample to make it scientifically representative showed that Kinsey was largely correct in this assertion.

The book made Kinsey nationally famous, and he traveled the country giving lectures. At the peak of his fame in 1949, he spoke to an audience of nine thousand in the field house at Berkeley—two thousand more than the record attendance at a basketball game. In 1950, the Institute for Sex Research moved into new quarters in the freshly renovated basement of ivy-covered Wylie Hall, one of the oldest buildings on campus. By 1951, the institute's budget had grown to over $100,000 (about $1 million in today's dollars) and the staff had expanded to fifteen.

Kinsey was now busy writing a second book, *Sexual Behavior in the Human Female*. The book, based on the sexual histories of nearly six thousand women, came out in September 1953. It reported the frequency with which women engaged in various types of sexual activity and how demographic factors affect patterns of sexual behavior. Kinsey discovered that unlike nearly all men, women could go for long periods without having sex. Most women were not aroused by pornography, fantasized less during masturbation and were much less likely to engage in sadomasochism, fetishes and transvestism. Kinsey found the female homosexuality rate was about half that of the male rate. It was a better, more scholarly book than the previous volume and easier to read, with fewer graphs and tables.

At first, the reaction was positive. All but one of the nation's leading magazines had favorable coverage (*Cosmopolitan* had an unfavorable reaction), as did nearly two-thirds of the leading newspapers. A Gallup poll found that about 75 percent of people thought that having this type of information available was a good thing. But then a furious backlash came from the pulpits of America. Billy Graham, who had not actually read the book but had read reports about the contents, said, "It is impossible to estimate the damage this book will do to the already deteriorating morals of the country." Kinsey's hometown newspaper, the Bloomington *Herald Telephone*, advised its readers to turn to the Bible. A New York politician even promised a congressional investigation. It was a sign of coming trouble.

The attacks didn't affect sales—the book sold much faster than the *Male* volume—but they did affect funding. Early in 1954, the Rockefeller Foundation ended its financial support of the institute. The reason was simple: they wanted nothing to do with sex research. The institute still had money coming in from book royalties, which was enough to keep it going for a few years. Kinsey needed a new source of income, but he refused to request government funding because that money came with strings attached, namely the right to inspect, which might compromise confidentiality.

The personal attacks on Kinsey took their toll, and after a tour of Europe in 1955, his health began to deteriorate. His heart, weakened by his childhood diseases, began to give out. While gardening one day, he hit his ankle, and a blood clot broke free. He was taken to the hospital and spent his last night in an oxygen tent. Alfred Kinsey died from a coronary embolism at 8:00 a.m. on August 25, 1956, at age sixty-two. His childhood doctor had predicted an early death at twenty-one, but Kinsey managed to live nearly three times that long. Kinsey is buried in Rose Hill Cemetery in Bloomington.

Many years after his death, Kinsey's own sexual behavior became known. He was bisexual and had sex with students, staff and collaborators. His wife was fully aware of this behavior and sometimes participated in the sexual activity. Indeed, Kinsey's own sexual complexity may have motivated him to take up sex research. Unfortunately, some have tried to use Kinsey's personal sexual conduct to discredit his work. But his own sexual behavior is irrelevant—it does not change the data.

When Wardell Pomeroy's biography *Dr. Kinsey and the Institute for Sex Research* was published in 1972, he revealed that Kinsey had also filmed sexual activity. If this had become publicly known at the time, it would have caused a scandal, so Kinsey and his team had kept it secret. Most of the filming took place in Kinsey's home in an attic bedroom furnished with a mattress. On one occasion involving violent sadomasochism, Clara Kinsey periodically entered the room and nonchalantly changed the sheets as the team watched and took notes.

Kinsey's main legacy is his documentation of the wide range and variety of human sexual experience. At the time of Kinsey's death, his team had collected approximately eighteen thousand sexual histories, of which Kinsey himself took nearly eight thousand. From extramarital sex to homosexuality to sex with animals, Kinsey calmly recorded everything. He claimed that in human sexuality, there is no normal and abnormal, just typical and

The Kinsey house at 1320 East First Street in the Vinegar Hill neighborhood of Bloomington, Indiana. *Wikimedia Commons.*

atypical. At a time when masturbation was a reason for being rejected by the U.S. Naval Academy, Kinsey showed that nearly all men and most women engaged in the practice and it didn't result in insanity, blindness or any other affliction. As Kinsey liked to say, "Everybody's sin is nobody's sin." Kinsey was neither a deep thinker nor a great theoretician; he didn't try to understand the psychology or emotion that accompanies sex. But as a collector of sexual data, he was unsurpassed, and the weight of that data eventually helped to change attitudes about human sexuality.

7

PERCY JULIAN

1899-1975

Field | Chemistry
Major Accomplishment | Synthesized important medical compounds
from plants, making them more affordable to mass produce
Indiana Connection | Attended DePauw University from 1916 to 1920,
taught at DePauw University from 1932 to 1936

> *I feel that my own good country robbed me of the chance
> for some of the great experiences that I would have liked to
> live through. Instead, I took a job where I could get one and
> tried to make the best of it. I have been perhaps a good
> chemist, but not the chemist that I dreamed of being.*
> *—from Julian's unfinished autobiography*

PERCY LAVON JULIAN was born on April 11, 1899, at the corner of
Jefferson Davis Avenue and South Oak Street in Montgomery, Alabama.
According to an episode of the PBS series *Nova*, Julian was the first of six
children born to James Sumner Julian, a railroad mail clerk, and Elizabeth
Lena Julian née Adams, a schoolteacher. Percy's grandfather had been
enslaved. The young boy grew up in the segregated South when Jim Crow
laws ruled the land. Julian's parents, both graduates of what would become
Alabama State University, saved enough money to create a small family
library for their children; the public library was closed to Black people.

At age twelve, Julian recalled picking berries in the woods and finding a
Black man hanging from a tree. He had been lynched a few hours earlier.

Julian said that the man didn't look like a criminal, he just looked scared. On his way back home he came across a rattlesnake and killed it. For many years after the incident, whenever Julian saw a white man, he would involuntarily see the contours of a rattlesnake. A reporter once asked what Julian's greatest nightmares were from growing up in the South. He answered, "White folks and rattlesnakes."

Public education for Black students only went up to the eighth grade in Montgomery. After that, Julian completed two years at the local teacher training school for African Americans. In 1916, with the equivalent of a tenth grade education, he headed north to DePauw University in Greencastle, Indiana. He entered as a "sub-freshman," meaning that in addition to his regular college courses, he had to take high school–level classes in the evening for two years to make up for his inadequate preparation. On his first day at college, Julian recalled that a white student welcomed him and stuck out his hand. "I had never shaken hands with a white boy before and did not know whether I should or not," Julian explained. "But you know, in the shake of a hand my whole life was changed. I soon learned to smile and act like I believed they all liked me, whether they wanted to or not."

Despite the warm welcome, campus life was segregated. Julian was not permitted to live in the dormitories. Instead, he was forced to stay in an off-campus boardinghouse with a slop jar as a toilet. When he asked what time dinner was served, the landlady told Julian that she was not expected to give him meals. It took him a day and a half just to find a place to eat. Eventually, he found a home in a fraternity house, where he waited tables, fired the furnace and did other odd jobs in exchange for meals and a bed in the attic. Overcoming all the obstacles, Julian finished at the top of his class, graduating in 1920 as a Phi Beta Kappa and as valedictorian. The entire Julian family eventually moved to Greencastle so that the other five children could go to DePauw. Julian's two brothers would go on to become doctors, and his three sisters would earn master's degrees.

Julian's favorite teacher at DePauw was William Blanchard, a chemistry professor who had a contagious enthusiasm for scientific discovery. Blanchard inspired Julian to pursue a doctorate so he too could do scientific research. But when Blanchard tried to get Julian accepted at Harvard, the head chemist replied that there was no point in applying because nobody was going to hire a Black researcher. The chemist suggested that Julian find a teaching job at a Black college in the South—a position that wouldn't require a doctorate. Julian reluctantly followed the advice and became a chemistry instructor at Fisk University, a historically Black college in Nashville. After two years, Harvard had a change of heart. Julian won a fellowship to the famous school where he earned a master's degree in 1923 with a straight A average. At that point, worried that white students would balk at being taught by a Black teacher, Harvard ended his teaching assistantship, making it financially impossible to continue toward a doctorate. So he took a teaching position at the West Virginia State College for Negroes. Unhappy there, he moved on to Howard University in Washington, D.C., the most prestigious Black college in America, where he taught for several years.

In 1929, Julian finally got a chance to earn his doctorate when he won a Rockefeller Foundation fellowship to the University of Vienna. At the time, chemists were fascinated by plants because they contained chemicals that affected people: coffee would keep you awake, tobacco would calm your nerves and foxglove could affect the heart. A major goal of chemistry in the early twentieth century was to understand what these natural chemicals were, identify their structure and figure out how to make them artificially. This was called natural products chemistry, and in 1929, Vienna was the subject's epicenter.

Julian studied under the Austrian chemist Ernst Späth, a giant in the field, who was particularly interested in a class of natural product chemicals called alkaloids because they seemed to be the most powerful. We now know that it's the alkaloid called caffeine in coffee that keeps you awake and the alkaloid called nicotine in tobacco that has a calming influence. Morphine, strychnine and cocaine are other plant-based alkaloids. For his doctoral research, Späth assigned Julian the task of finding one of these alkaloids and figuring out its chemical structure. It was a difficult problem that pushed Julian to the limits of his ability, but he met the challenge. He found the chemical, identified its structure and, in 1931, earned his doctorate in chemistry, only the second African American to do so. When Späth was asked about Julian, he replied that he was "an extraordinary student the likes of which I have never had before in my career as a teacher."

Outside of the lab, Vienna and its surroundings beckoned. Europe was largely free of the racial attitudes and prejudices that prevailed in the United States. There were no segregated buses, drinking fountains or restrooms. In Europe, Black people were a novelty, and Julian reveled in this new world. Julian quickly learned to speak German and enjoyed an active social life. He skied in the Alps, swam in the Danube, played tennis, took piano lessons and attended the opera. Julian's time in Vienna greatly influenced his personality and his future scientific career.

A self-confident Julian returned to Howard University as chairman of the chemistry department and brought with him Josef Pikl, a fellow chemistry student at Vienna. But the job lasted only a year because Julian found himself embroiled in some nasty and embarrassing university politics. At the request of the university president, Julian goaded a white chemistry professor, Jacob Shohan, into resigning. In retaliation, Shohan handed a stack of letters that Julian had written him from Vienna over to the local African American newspaper. In addition to discussing his romances, the letters also contained some unflattering comments about certain members of the Howard faculty, calling one dean an "ass."

If that weren't enough, Julian also got into a personal dispute with his laboratory assistant, Robert Thompson, whom Julian recommended for termination. Thompson retaliated by suing Julian for having an affair with his wife, Anna Roselle Thompson. Julian countersued for libel. When Thompson was fired, he too supplied the local newspaper with letters from Julian. In the letters, Julian boasted about how he fooled the university president into agreeing to his plans for the chemistry building and how he bluffed a friend into appointing a professor that Julian liked. Throughout the summer of 1932, all of Julian's letters were published in the *Baltimore Afro-American.* The scandal resulted in Julian's resignation.

Julian was rescued by his old professor William Blanchard, who offered him a position as a research assistant teaching organic chemistry at DePauw. Julian happily took the job, and again, Pikl followed. Soon, Julian, Pikl and DePauw students were sending scientific papers to the nation's leading chemistry journals. A total of eleven papers were published during Julian's tenure at DePauw, a big number for such a small school. Julian put DePauw on the map for undergraduate research. A former student recalled that as a teacher, Julian was something of a showman. He would come in to the hall with a flourish wearing a white lab coat and give his lecture with great oratorical skill.

Julian's big breakthrough came in 1935, when Julian and Pikl succeeded in chemically synthesizing physostigmine, an alkaloid found in the African Calabar bean. (Synthesis is the process of making a natural product artificially in the lab from simple building blocks.) Physostigmine was useful in treating glaucoma. By making it in the lab rather than coaxing it from the bean, larger quantities could be made. Julian and Pikl had some formidable competition. One of the most famous chemists in the world, Sir Robert Robinson, was also working to make the chemical. Much to Julian's dismay, Robinson seemingly got there first and published a paper on his method.

But Julian found a mistake in the paper that implied that Robinson had found the wrong chemical. In his own paper, Julian described a simpler and successful method. In 1999, the American Chemical Society designated Julian's work on physostigmine a National Historic Chemical Landmark, saying it was "the first of Julian's lifetime of achievements in the chemical synthesis of commercially important natural products." A commemorative plaque can be found at DePauw.

Julian also had some personal good news in 1935. On Christmas Eve of that year, he married Anna Roselle Johnson, the woman with whom he was accused of having an affair at Howard. Anna had divorced Robert Thompson and was working on a doctoral degree in sociology. For the first three years of the marriage, she lived in Washington to work in the public schools and finish her education. In 1937, she became the first African American woman to earn a doctorate in sociology. The couple would go on to have two children: Percy Lavon Julian Jr., who became a civil rights lawyer, and Faith Roselle Julian, who gave speeches about her famous father.

Julian's synthesis of physostigmine made him a major figure in the world of chemistry, but it wasn't enough to secure his position at DePauw. After members of the American Legion were fired from the university, the group retaliated by complaining to the school about "hiring a negro, who had been discharged from a government school faculty, and an Austrian, who faced deportation proceedings." The local newspaper ran a headline reading "DePauw Trustee Board White Washes His Errors." Julian was denied a professorship and forced to leave.

Julian decided to seek a job in the private sector. DuPont offered a job to Julian's colleague Joseph Pikl but declined to hire Julian because they were unaware he was Black. Julian insisted that Pikl take the job; he did and spent the rest of his career there. Julian applied to numerous companies, but their response was always along the lines of "we've never hired a Black research chemist before…we're not sure how it would work out."

Julian thought he had a job with the Institute of Paper Chemistry. Unfortunately, the institute was located in Appleton, Wisconsin, a sundown town where the law stated: "No Negro should be bed or boarded overnight in Appleton." However, William J. O'Brien, vice president of Glidden, sat on the board of the institute and had been looking for a smart chemist to run the company's new facility in Chicago. O'Brien offered Julian the job at Glidden's Soya Products Division—not as a member of the research team, but as the director of research. At the time, ten years before Jackie Robinson broke the color barrier in baseball, this was unheard of.

Julian accepted the job, moved to Chicago and focused his attention on the soybean, a plant that the president of Glidden thought would be integral to the company's future. Julian's first task was to isolate the protein in soybeans, and he drove his team of chemists hard to find the compound. They did, and it turned out to be useful in products such as paper coating and latex house paint. Then the team tore the bean apart to find uses for its other chemicals. This led to dozens of new products, including smoother chocolate, new salad oils and non-spattering margarine. Soybean meal was used in plastics, linoleum, plywood glue, livestock feed and dog food. Under Julian's demanding leadership, Glidden's new soybean division was a success.

Julian was reunited with his wife, and they settled into a home in Maywood, a suburban community on the west side of Chicago. Julian began to get restless regarding the type of practical industrial chemistry he was doing. Instead, he yearned to do basic scientific research that got at the heart of nature.

Julian was fascinated by a hormone called progesterone, partly because his wife had suffered from several miscarriages. Discovered in 1934, progesterone was one of a class of compounds called steroids, which scientists were beginning to understand played key roles in reproduction, growth, sexual development and injury response. Despite the variety of functions, the steroids had similar chemical structures. Progesterone was known as the "pregnancy hormone" because it played an important role in preparing the uterus for childbirth. In the 1930s, nearly one out of six pregnancies ended in miscarriage or premature birth, resulting in the loss of hundreds of thousands of babies every year. At the time, steroids could be extracted in tiny amounts from vast volumes of horse urine and hundreds of pounds of animal spinal cords. A breakthrough came when scientists realized that plants also have steroids whose chemical structures were very similar to that of animal steroids. Beginning with the plant steroids, chemists might be able to synthesize large quantities of human steroids. Julian and other chemists began to look for ways to accomplish this synthesis.

In 1938, a chemist named Russell Marker found a way to convert sarsaparilla root into progesterone, but the method was too expensive to be practical. A cheaper source was needed. Julian discovered such a method purely by accident. One day, he got a frantic call informing him that a tank containing 100,000 gallons of refined soybean oil was spoiled because water had leaked into the tank. He ran over to inspect the tank and found a white sludge at the bottom. His despair turned into ecstasy when he found crystals in the sludge. Remembering his work with the Calabar bean at

DePauw, Julian knew the crystals were stigmasterol, a plant chemical that could easily be converted into progesterone. The stigmasterol had been forced out of the soybean oil by water. Julian then developed an industrial process by which progesterone and other steroids could be produced from stigmasterol in large quantities. Glidden was now a major player in the human sex hormone business. In 1940, via an armed guard, Julian sent a one-pound package of progesterone valued at $63,500 to the Upjohn Pharmaceutical Company. It was the first commercial shipment of an artificial sex hormone anywhere in America.

Julian's discoveries were making him famous. In 1947, the NAACP presented him its highest award for helping develop a firefighting foam used during World War II on aircraft carriers. It saved thousands of lives. Julian was featured by the *Reader's Digest* in an article titled "The Man Who Wouldn't Give Up." He was appointed to the board of half a dozen major universities, was named Chicagoan of the year and was in demand as a public speaker.

By 1950, the Julians, now with two children, had outgrown their home in Maywood and moved to a new home in the affluent community of Oak Park. While most Oak Park residents welcomed the new residents, a few did not. Before the Julians could even move in, an attempt was made to fire-bomb the house. His ten-year-old son later recalled seeing jars of gasoline and smelling fumes in the house. A cloth fuse ran from outside the house to the jars inside. Luckily, the arsonists made the mistake of closing the door. When the fuse was lit, the door formed a seal keeping the flames out. When the son asked his mother why someone had done this, she explained that there were people who didn't want them to live there because of the color of their skin.

But the Julians could not be intimidated. Things calmed down for a while, but one night in June 1951, the children were home alone with a babysitter when vandals attacked the house with dynamite. Again, the attack was unsuccessful and merely blew a hole in the ground. But the incident enraged Julian, and for several nights, he sat outside in a tree with his son, armed with a shotgun. The son later joked that at least he got to spend some quality time with his dad. The majority of the residents of Oak Park were shocked and outraged by the attacks, apologized to the Julians and organized a march to show their support.

Percy Julian weathered the storm and set off to tackle a new scientific challenge. Rheumatoid arthritis, an inflammation of the joints, was a painful and crippling disease that affected millions. Scientists had been seeking a cure for centuries, but all they had come up with were dubious treatments such as mineral baths and cobra venom. Finally, in 1949, Philip Hench of the Mayo Clinic announced at a rheumatology conference the dramatic effectiveness of a chemical called Compound E, which later became known as cortisone. He showed before and after film footage of patients who had been treated with the drug—patients who moved only with difficulty could, after treatment, run and jump. At the end of the talk, the audience stood up and cheered. It was a scientific miracle. The problem was that it could only be produced in minute quantities from the bile of cattle. And the process by which it was synthesized was incredibly complicated, requiring more than thirty steps. To make it practical, a more plentiful starting material was required along with a simpler synthesizing process. Chemists, including Julian, scurried off to their labs to find a procedure that would work.

Cortisone had a chemical structure similar to steroids, and Julian, already familiar with the chemistry, hoped to make it from soybeans. He was able to produce a substance called Reichstein's Substance S, which was different from cortisone by a single missing oxygen atom. But inserting that one little oxygen atom in the right position proved to be extremely difficult. Julian thought that if he made enough Substance S, he or someone would eventually figure out how to do it. He continued to struggle with the problem for two years. Meanwhile, the cost of cortisone topped $4,000 per ounce, one hundred times the price of gold.

Finally, in April 1952, chemists at Upjohn in Kalamazoo, Michigan, announced they had discovered a common mold that could effortlessly insert the oxygen atom into the required position. Start with a tank of Substance S, dump in the mold and fish out the cortisone. This marked the end of the cortisone crisis. But there was another starting material, progesterone made by Syntex, that was cheaper. Progesterone won the day, and Glidden decided to get out of the unprofitable steroid business altogether. Glidden sold the license for Substance S to Pfizer and Syntex, and Julian was ordered to teach their chemists how to make it using the process that Julian had invented. Soon, the situation became untenable for Julian. In December 1953, at age fifty-four, after eighteen years of service and 109 patents, he left Glidden. He also left behind a generous $50,000 salary (the equivalent of nearly $500,000 in today's money).

Julian immediately started his own company, Julian Laboratories, housed in a small concrete-block building infested with mice and rats in Franklin Park, Illinois. He brought with him some of the chemists he had worked with at Glidden, including African Americans and women. The company would focus on the manufacture of steroid intermediates, compounds that were one step from a finished steroid product. The plan was to make them faster and cheaper and then sell them to the major pharmaceutical companies. He quickly landed a contract from Upjohn for $2 million, and more work followed from Pfizer, Merck and others.

Julian's little company struggled to survive in a highly competitive market. Shipments had to be made—the financial penalties for missing a shipment were severe. Former employees tell of working into the wee hours of the morning. Julian told one employee that sleep could be bad for his health—that he could die in his sleep—and he should get back to work. But the employees were unusually loyal to the company and wanted to see it succeed, so they put up with the long hours and working conditions. Eventually, the company made Julian a millionaire and one of the richest Black businessmen in America. The company employed more Black chemists than any other company in the country, and many of his employees used their jobs at Julian Laboratories as a springboard for positions in business and academia.

Julian continued to fight against the racial conditions of his time. At meetings of the American Chemical Society, he took the arm of a colleague to go into a meeting so others would know he belonged. As he grew older, he set aside science and worked toward racial justice. He battled against racial discrimination in jobs and housing, raised funds to aid civil rights lawyers and preached that education was the key to the advancement of the race. During the 1960s, he began to see that the old ways of serving as a model citizen and patiently waiting for change was not working and started supporting the more confrontational tactics of the younger generation. His son went to Nashville to support the desegregation of lunch counters. Julian was proud but, at the same time, afraid. He warned his son that this was not a game and that these people were for real. "So are we," his son responded.

As he entered old age, Julian became a grandfather. One of his granddaughters

Postage stamp commemorating Percy Julian. *Shutterstock.*

recalled being shown a doll that had been sent to Julian by a woman who had arthritis and couldn't use her hands. After taking cortisone, she was able to knit the doll. She sent the doll to Julian as a token of her appreciation for the role he played in finding a cure. Meanwhile, Julian's friend Bernhard Witkop began a quiet campaign to try to get Julian elected to the National Academy of Sciences, one of the highest honors for an American scientist. It wasn't easy, but Witkop persisted. In 1973, Julian got an unexpected call from the academy informing him of his election. He was only the second African American elected to the distinguished group.

In April 1975, a week after his seventy-sixth birthday, Percy Julian died of cancer. He is buried in Elm Lawn Cemetery in Elmhurst, Illinois.

8

WENDELL STANLEY

1904-1971

Field | Biochemistry
Major Contribution | Won the Nobel Prize for Chemistry in 1946 for
   his work on viruses
Indiana Connection | Born in Ridgeville and graduated from Earlham
   College

> *The viruses hold the key to the modification—for better
> or worse—of all life. They hold the key to the secret of
> life, to the solution of the cancer problem, to biological
> evolution, to the understanding and control of heredity,
> perhaps to the nature of all future life on earth.*
>   —*Stanley in 1956, quoted in his* New York Times *obituary*

WENDELL MEREDITH STANLEY was born in Ridgeville, a small town in
east-central Indiana, on August 16, 1904. His parents, James Stanley and
Claire Plessinger, published two local newspapers, the *Ridgeville News* and the
Union City Eagle. When James died in 1920, the family moved to Richmond.
Wendell graduated from Richmond High School in 1922 and enrolled at
Earlham College, also in Richmond, where a family ancestor had donated
land for the college with the provision that anyone bearing the Stanley family
name would be given special consideration. Stanley majored in chemistry
and mathematics and was captain of the football team. After earning his
degree in 1926, his plan was to become a football coach, but after meeting

with Roger Adams, head of the chemistry department at the University of Illinois, he changed his mind.

He began graduate work in organic chemistry at Illinois, where he earned his doctorate in 1929. He married Marian Staples shortly after graduating on June 15, 1929. The couple eventually had four children, a son and three daughters. Stanley stayed at Illinois for another year as a postdoctoral research assistant, then left as a National Research Council Fellow to do research in Munich, Germany, with Heinrich Wieland, a Nobel laureate in chemistry. In late 1931, he returned to the United States and accepted a position as an assistant at the Rockefeller Institute for Medical Research in New York City. In 1932, Stanley transferred to the institute's Department of Plant and Animal Pathology, located in Princeton, New Jersey. It was there that Stanley was introduced to the field of virology, where he would do his most important work.

At the time, scientists were baffled by viruses. They were known to cause contagious diseases, but nobody knew exactly what they were. Some thought they were incredibly tiny living organisms too small to be seen under a microscope. Others thought they were strange but lifeless chemicals. Working in his Princeton laboratory, Stanley solved the problem.

Stanley's research focused on the tobacco mosaic virus, an infection that caused unusual mosaic-like patterns and discolorations on tobacco leaves, ultimately killing the plants and devastating crops. To isolate the virus, Stanley ground up a ton of infected tobacco plants and painstakingly purified the extracted juices. He ended up with about a tablespoonful of a white powdery substance—an almost pure sample of crystallized tobacco mosaic virus. The virus could sit on a shelf, dry and inert, for years. Then, when dissolved in water and placed on a tobacco leaf, the chemical would spring back to life, invade cells, reproduce millions of times and spread to other plants. Delving further into the chemical structure of the virus, Stanley showed that it was composed of protein and ribonucleic acid (RNA).

Stanley discovered that viruses occupy a strange twilight zone between the living and the nonliving. The finding created a scientific sensation, with some comparing it to Pasteur's discovery of the role of bacteria in causing disease. For his discovery, Stanley shared the 1946 Nobel Prize for Chemistry with John H. Northrup, a colleague at the Rockefeller Institute who also worked in virology, and James B. Summer of Cornell University.

During World War II, the military was plagued by influenza epidemics. Stanley was asked if he could develop a vaccine to prevent the illness. Eager

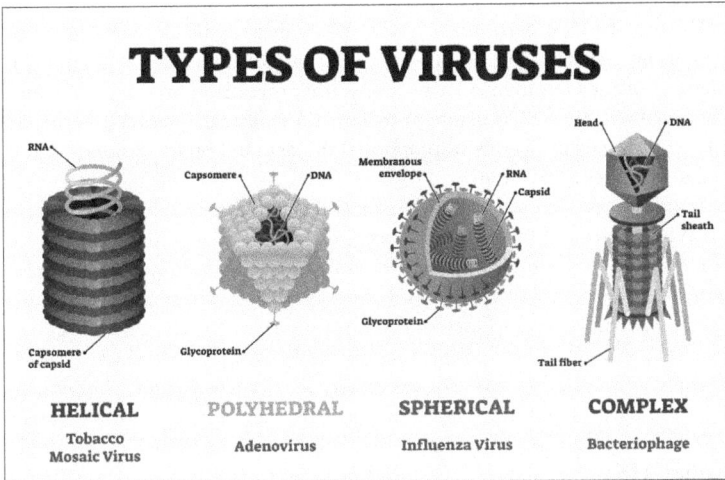

TYPES OF VIRUSES

HELICAL
Tobacco
Mosaic Virus

POLYHEDRAL
Adenovirus

SPHERICAL
Influenza Virus

COMPLEX
Bacteriophage

Four main types of viruses. *Shutterstock.*

to help the war effort, he went to work, successfully purified some of the most common influenza viruses and produced a vaccine that was partly effective. Stanley's war work paved the way for modern flu vaccines that each year prevent millions of illnesses and save thousands of lives.

In 1946, Stanley was on his way to the University of California to accept an honorary degree when his plane was grounded for several hours in Wyoming due to fog. He learned that the president of the University of California, Robert Gordan Sproul, was on the flight and the two men struck up a conversation. Stanley talked to Sproul about the promising future of virus research. Sproul was so impressed that he later asked Stanley to come to Berkeley and establish and direct a virus laboratory. At about the same time, the Rockefeller Institute announced it was closing its Princeton facility, so Stanley decided to accept Sproul's invitation, and in 1948, he became a professor of biochemistry at Berkeley.

At the new laboratory, Stanley recruited an outstanding team of scientists, directed a series of important research achievements and trained some of the world's most promising students. In 1954, Stanley and his colleagues made yet another discovery that helped save millions of lives: they crystallized the polio virus. This was the first step toward understanding the structure of the virus and developing a vaccine against it.

During the McCarthy era of the 1950s, university faculty were asked to sign loyalty oaths to the United States, which stated, in part, that "I am not a member of the Communist Party or any other organization which advocates the overthrow of the Government by force or violence." Stanley was one of a large group of professors who opposed the demand on the grounds that it violated academic freedom.

Stanley was a lifelong Republican and headed the statewide "faculty for Nixon" effort in 1962. However, he had no use for politicians who failed to support science and higher education with adequate funding—a policy he believed was foolish and shortsighted. In an interview, Stanley cited dwindling state support for research as part of the reason he resigned as director of the virus laboratory in 1969. In his later years, Stanley devoted much of his time to the American Cancer Society, helping to raise funds and organize conferences. He was convinced that cancer was caused by a virus and that a cure might be found through virologic studies. He was also an advisor and consultant to the U.S. government and the World Health Organization.

Stanley died of a heart attack on June 15, 1971, at age sixty-six while traveling in Spain. In a tribute, a colleague described him this way: "Stanley had the skills, the intelligence, and the driving determination of a great scientist....But there was also an easy affability about the man, a kindliness and warmth (a 'country doctor' quality, some have said) that made him at home with people from all walks of life. He was always patiently willing to explain and interpret and teach."

9

EMIL KONOPINSKI

1911–1990

Field | Physics
Major Contribution | Worked on the Manhattan Project, the World War II effort to build an atomic bomb
Indiana Connection | Born in Michigan City and was a professor at Indiana University from 1938 until 1990

> It is shown that, whatever the temperature to which a section of the atmosphere may be heated, no self-propagating chain of nuclear reactions is likely to be started. The energy losses to radiation always overcompensate the gains due to the reactions.
> —from a declassified paper coauthored by Konopinski regarding the possibility of igniting the atmosphere with nuclear bombs

EMIL JAN KONOPINSKI was born in Michigan City, Indiana, on Christmas Day in 1911. His parents, of Polish descent, were Joseph Konopinski and Sophia Sniegowska. He attended the University of Michigan, where he earned his doctorate in 1934 in theoretical physics with an expertise in beta decay, one type of radioactivity. From 1936 until 1938, Konopinski was a National Research Council fellow and worked on nuclear reactions with physicist Hans Bethe, a Nobel laureate at Cornell. In 1938, he joined the physics department at Indiana University, where he and other faculty members began a program in nuclear physics.

During World War II, he worked on the Manhattan Project, the effort by the United States to build an atomic bomb. He joined the University of Chicago's Metallurgical Laboratory, where he worked under the great Italian American physicist Enrico Fermi. (The Metallurgical Laboratory or "Met Lab" was a name given to the lab to hide what they were really working on.) He arrived at the Met Lab at about the same time as Edward Teller, later known as the "father" of the hydrogen bomb. Teller recalls in his memoir, "We were newcomers in the bustling laboratory and for a few days we were given no specific jobs." So Teller proposed that he and Konopinski check his calculations that seemed to show that using an atomic bomb to ignite a thermonuclear explosion in deuterium (in other words, a hydrogen bomb) was impossible. But as they worked through the equations, they realized that a hydrogen bomb *was* possible.

In the summer of 1942, Robert Oppenheimer, head of the top secret Manhattan Project, called together an elite corps of theoretical physicists at Berkeley. He called the small group the "luminaries" because their job was to shine light on the actual design of an atomic bomb. Konopinski and Teller were called out to the secret meetings, held in Oppenheimer's office in the northwest corner of the fourth floor of LeConte Hall. The meetings, more like seminars, lasted for several weeks, with the scientists arguing and debating about the physics of the bomb. They also discussed the possibility of building the hydrogen bomb that Teller and Konopinski had shown was theoretically possible. At one point, Konopinski had an idea, or a "conversational guess" as Teller would later call it. Teller says in his memoir, "Konopinski suggested that, in addition to deuterium, we should investigate the reactions of the heaviest form of hydrogen, tritium." Including tritium would reduce the ignition temperature of the reaction. Ultimately, it was decided that the atomic bomb would be built first, but Konopinski's suggestion would later be used in the development of the hydrogen bomb.

Another problem that Teller brought up to the assembled luminaries was the possibility the bombs might ignite the earth's atmosphere or its oceans and destroy the entire planet. Konopinski wrote a top-secret paper (now declassified) with Teller and Cloyd Marvin Jr. showing that the explosion of nuclear weapons would *not* set off a chain reaction in the atmosphere. Konopinski wound up working at Los Alamos, the top-secret laboratory in New Mexico where the atomic bomb was designed and built, until the end of the war.

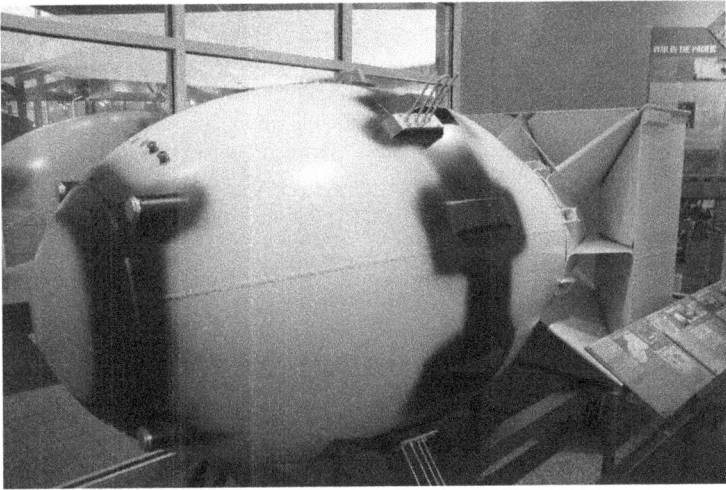

Replica of the bomb dropped on Nagasaki at the end of World War II at the Bradbury Science Museum, Los Alamos, New Mexico. *Jeffrey M. Frank/Shutterstock.*

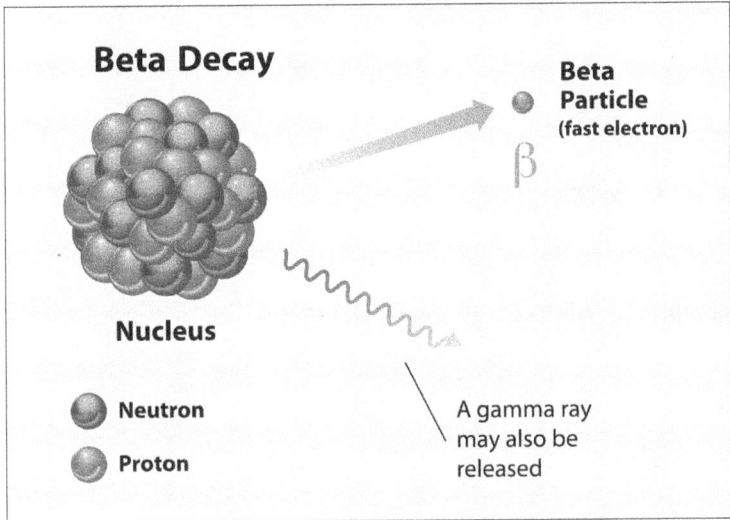

Radioactive beta-decay where a neutron changes into a proton and an electron. The proton remains in the nucleus and the electron (beta particle) is ejected. *Shutterstock.*

After the war, Konopinski returned to Indiana University to teach and do research on nuclear structure, nuclear reactions and the weak interaction. Working with his students, he made major contributions to the understanding of beta-decay experiments, the concept of lepton conservation and the relationship between nuclear models and beta-decay. This culminated in a 1966 book titled *The Theory of Beta Radioactivity*. He continued to be somewhat involved in the development of the hydrogen bomb, attending a conference of about thirty scientists in April 1946 and serving on the twenty-five member "Family Committee" in 1950. He also served as a consultant to the Atomic Energy Commission from 1946 to 1968.

Emil Konopinski died at age seventy-eight of a cardiac arrest on May 26, 1990. In an obituary, colleagues remembered him as a "deeply humble person who avoided any occasion to judge others."

HERBERT BROWN

1912–2004

Field | Chemistry
Major Contribution | Won the Nobel Prize for Chemistry in 1979 for developing important reagents in organic synthesis
Indiana Connection | Professor at Purdue University from 1947 until his death in 2004

> *Why did I decide to undertake my doctorate research in the exotic field of boron hydrides? As it happened, my girlfriend, Sarah Baylen, soon to become my wife, presented me with a graduation gift, Alfred Stock's book,* The Hydrides of Boron and Silicon. *I read this book and became interested in the subject. How did it happen that she selected this particular book? This was the time of the Depression. None of us had much money. It appears she selected as her gift the most economical chemistry book ($2.06) available in the University of Chicago bookstore. Such are the developments that can shape a career.*
> *—Herbert C. Brown in his Nobel lecture, December 8, 1979*

HERBERT CHARLES BROWN was born in London, England, on May 22, 1912. His parents, Charles Brovarnik and Pearl Gorinstein, were born in the city of Zhitomir in northwestern Ukraine and had come to London in 1908 as part of a wave of Jewish immigration in the early 1900s. They were married in London and lived in an apartment complex built to house immigrants. Charles was a cabinetmaker, doing delicate inlaid work. Herbert was the couple's second child—a sister, Ann, had been born in 1909.

In June 1914, Charles decided to join his parents and other family members in Chicago. Charles's parents had anglicized the family name to Brown, so the new arrivals also took that name. The couple had two more children: daughters Sophie in 1916 and Riva in 1918. In 1920, Charles decided to go into business for himself. He opened a hardware store, and the big family squeezed into a little apartment above the business.

Herbert did well in school and skipped several grades, although he refused to advance into the same class as his older sister. At age twelve he entered Englewood High School on Chicago's south side. Unfortunately, in 1926 Charles died from an infection, and Herbert had to drop out of school so he could help run the hardware store. "I am afraid that I was not really interested in the business and spent most of my time reading," Brown later recalled. After three years, his mother finally decided that she would mind the store so Herbert could return to school. He reentered Englewood in 1929 and graduated in 1930.

The family sold the hardware store the year Herbert graduated. For the next several years, Herbert had a series of temporary jobs, but nothing permanent; it was the Great Depression, and jobs were scarce. With nothing else to do, he decided to go to school and entered Crane Junior College in February 1933. He was planning on majoring in electrical engineering because he had heard that you could make a good living in that field. Then he took a chemistry class, was fascinated by the subject and decided instead on a career in chemistry. It was at Crane that Brown met his future wife, Sarah Baylen. Sarah had been the best chemistry student at Crane until Brown came along. She later admitted that initially, she "hated his guts." Brown recalled, "But since she could not beat me, she later decided to join me, to my everlasting delight."

The semester after Brown enrolled, the school closed due to a lack of funding. So Brown went to night school at the Lewis Institute, where he took a couple of classes while working during the day as a part-time shoe clerk. Then he heard that one of his professors at Crane, Dr. Nicholas Cheronis, was inviting some of his former students to do experimental work in a small commercial lab located in Cheronis's garage. Brown and Baylen jumped at the opportunity. When Wright Junior College opened its doors in 1934, Cheronis was hired to teach chemistry, and Brown and Baylen enrolled as students. The pair graduated in 1935 with associate degrees, and Baylen prophetically signed Brown's yearbook with the words "To a future Nobel Laureate!"

Brown then took the competitive examination for a scholarship to the University of Chicago. He won a partial scholarship and entered the University of Chicago in the fall of 1935. At the time, the university's president, Robert Maynard Hutchins, supported the idea of students proceeding at their own pace. Thus, a student could take more than the usual three courses per semester for the same cost. So Brown signed up for an unbelievable ten courses. This enabled him to finish his junior and senior years in just three quarters, earning his bachelor's degree in 1936. As a graduation present, Baylen bought her boyfriend a book titled *Hydrides of Boron and Silicon* by Alfred Stock.

Initially, Brown did not intend on pursuing graduate work—he wanted to find a job and get married. But a famous chemist at the university, Julius Stiglitz, called Brown into his office and told him that he had a bright future as a research chemist, but such a position required a doctorate. Brown discussed the idea with Sarah, and she agreed that marriage could wait, so Brown began work on his doctorate. Brown had read the book Sarah had given him and became interested in the subject. As luck would have it, Professor H.I. Schlesinger's lab at the University of Chicago was one of only two labs in the world working on the hydrides of boron, so Brown decided to work under Schlesinger.

Marriage couldn't quite wait until Brown had his degree, and the couple were married "secretly" on February 6, 1937. The secrecy was due to the fact that marriage was against the advice of Stiglitz. What the lovebirds did not realize was that marriages were announced in the newspaper, so their secret lasted only a weekend. Brown was paid $400 annually as a graduate assistant, but tuition was $300. To make ends meet, Sarah got a job in medical chemistry at a local hospital. The couple would go on to have one son, Charles A. Brown, who followed in his father's footsteps and became a chemist.

Brown earned his doctorate in 1938 and tried to get a job in an industrial lab but was unsuccessful. So he accepted a postdoctoral position with Professor Morris Kharasch at the University of Chicago, where he did research on chemical reactions involving chlorine. A year later, Brown was back with Schlesinger's group as a research assistant working mainly on reactions involving lithium and sodium borohydrides. He was also a chemistry instructor.

At that time, tenure was not granted until after ten years. Brown had witnessed several young chemists remain at Chicago for nine years only to be denied tenure and then have to frantically search for another position. Brown wanted to avoid that fate, so he asked Schlesinger about his future with the department. Schlesinger told him that he had no future, so Brown sought another position.

As it turned out, Morris Kharasch had a friend, Neil Gordon, who had gone to Wayne University (now Wayne State University) in Detroit as the head of the chemistry department. Kharasch convinced Gordon to hire Brown. So in 1943 at age thirty-two, Brown moved to the Motor City as an assistant professor. Gordon was trying to establish a doctoral program at Wayne, but the laboratories were poorly equipped. Determined to do research, Brown had to find projects that didn't depend on fancy and expensive equipment. He decided to develop empirical theories that attempted to explain poorly understood steric effects, interactions that govern the shape and reactivity of ions and molecules. In 1946, Brown was promoted to associate professor.

Brown's pioneering work at Wayne established him as a productive researcher. Word got out, and in 1947, Brown was offered a position as a full professor in the chemistry department at Purdue University, where he stayed for the remainder of his career. In 1978, he retired and became an emeritus professor, although he continued to work with postdoctoral students. At Purdue, Brown continued his research on steric effects and resumed his work on the chemistry of borohydrides and aluminohydrides. Brown joked that his parents had the foresight to give him the initials H, C and B, which correspond to the chemical symbols of the three elements involved in hydroboration: hydrogen, carbon and boron.

In 1979, Brown shared the Nobel Prize in chemistry with Georg Wittig, a German chemist. The Nobel Prize citation reads, "For their development of the use of boron- and phosphorus-containing compounds, respectively, into important reagents in organic synthesis." To understand what this means, let us first consider the word *synthesis*. In chemistry, molecules can be artificially built or synthesized, even if they do not exist in the natural world. This is especially true in organic chemistry—chemistry involving carbon atoms—where the possibilities for new molecules is almost unlimited. A reagent is a substance that is added to a system that causes a reaction to occur. So what Brown did was to create new reagents containing the element boron. These reagents can be used to synthesize new organic compounds. For example, one of Brown's reagents is sodium boro-hydride ($NaBH_4$), which can become the preferred reagent for the reduction (the gaining of electrons by one of the atoms in the reaction) of carbonyl compounds (compounds where a carbon atom is double bonded to an oxygen atom). In addition, Brown created an entirely new class of compounds called organoboranes, which have become the most versatile reagents in organic chemistry. This

Gate at Purdue University in West Lafayette, Indiana. *Shutterstock.*

has resulted in new ways of connecting carbon atoms together. Wittig did essentially the same thing using phosphorus.

Brown had the kind of logical and rational mind that characterizes the best scientists and had a knack for avoiding incorrect conclusions. He served as a mentor to over three hundred advanced students, many of whom hold faculty positions in universities around the world. One of the people Brown influenced was the Nobel Prize–winning chemist and Purdue professor Ei-ichi Negishi (also portrayed in this book), who wrote that the single most important lesson he learned from Brown was to "live with eternal optimism." Negishi describes working with Brown in this way: "He would openly and repeatedly tell his group members that he did not believe in the value of spending much more than eight hours a day in the laboratories. Instead, he emphasized the significance of good thinking, planning, preparation, execution of laboratory experiments in a highly efficient manner, and timely interpretation of the results. In short, he expected each of us to be a successful entrepreneur, working jointly with him."

Herbert C. Brown died on December 19, 2004, after a heart attack. He was ninety-two. His beloved wife, Sarah, who Brown always credited with taking care of all the details of their personal lives so he could concentrate on science, died just a few months later. Brown once joked that after receiving the Nobel Prize, he carried the medal while she was the guardian of the check. Brown is buried in the Jewish Cemetery in Lafayette, Indiana.

II

SALVADOR LURIA

1912–1991

Field | Biology

Major Contribution | Won the Nobel Prize for Medicine or Physiology in 1969 for discoveries about the replication mechanism and the genetic structure of viruses

Indiana Connection | Professor at Indiana University from 1943 to 1950

> *What I have liked most in science has been the problem-solving activity and the sense of order that this activity generates. I like seeing patterns emerge, answers dovetail to create an intellectually simple and satisfying picture.*
> —*Salvador Luria in* A Slot Machine, A Broken Test Tube:
> An Autobiography, *114*

SALVADOR EDWARD LURIA was born into a middle-class Jewish family on August 13, 1912, in Turin, an industrial city in northern Italy at the foot of the Alps. His father, David Luria, was an accountant and managed a small printing business. His mother, Esther Sacerdote, had the equivalent of a junior high school education, but she was extremely well-read. Salvador also had a brother, Giuseppe, who was six years older.

As a child, Luria remembered his mother as being smart and attractive. But when he was in elementary school, a near fatal illness left her a semi-invalid who became dependent on opiates. She became obsessed with the family's health, convinced that some deadly disease was just around the corner. According to Luria, this created a home environment where sickness was rewarded. The main consequence of this strange familial hypochondria was that Luria became totally inept at sports or physical exercise. He tried skiing and tennis but found himself clumsy and unable to learn. As he grew older, he no longer felt obligated to participate in sports and even the smell of a gymnasium made him nauseated. This left Luria with a fear of situations where physical violence was a possibility. But as we shall see, Luria could never be accused of being a coward in the intellectual arena.

Luria did well in school, with the surprising exception of his science classes, which he nearly failed. He attributes this to the fact that the classes emphasized rote learning and vocabulary and he had no practical need for all the verbiage. Among his best friends was Ugo Fano, the son of a mathematics professor, who, according to Luria, "more than anyone else encouraged and helped me to become a scientist." Hanging out with Ugo and his intellectual family influenced Luria to pursue an academic life.

Luria went to school at a time when Fascism was restraining political freedom. He attended one of Turin's best high schools, where several teachers were actively fighting against Fascism and would eventually become leaders in post-Fascist Italy. Luria's most influential teacher was Augusto Monti, a writer and literature teacher who would later spend several years in jail for his anti-Fascist activities. But Monti's impact in the classroom was personal, not political. Luria recalled, "One felt confronted by a personality of deep intellectual and human integrity, a teacher who demanded and extended respect and who treated adolescent students as equals in rights if not in knowledge." Luria remembered when Monti's eyes filled with tears as he read poems about freedom to a class of impressionable fifteen-year-olds. "Acquaintance with young anti-Fascists," Lurie later claimed, "made it clear to me that it was respectable to risk jail for an idea."

After high school, Luria decided to go to medical school at the University of Turin, a path that pleased his parents but he was ambivalent about. His medical school years were difficult; an economic depression hurt his family's finances, and his mother was chronically ill. Luria persevered with the help of Ugo Fano, who had decided to pursue a career in physics. Ugo taught Luria calculus and told him of the revolutionary discoveries in quantum physics. Soon, Luria began dreaming of a career as a research scientist.

Luria finished medical school in 1935 near the top of his class but had little desire to actually practice medicine. Instead, he decided to search for a medical specialty closely related to physics and use it as a bridge to the field of biophysics. The obvious choice was radiology, a medical specialty that, at the time, used X-rays to image the body. So he joined the radiology department in Turin and signed up for a class. But Luria was profoundly disappointed by his experience with radiology, finding the field dull and the class a farce—the professor couldn't even write a simple equation correctly.

Luria was saved from radiology by, of all things, the army. At the time, doctors in Italy were required to spend eighteen months in military service—five in training and thirteen as junior medical officers. Luria could have easily been exempted from service because of his poor physical condition and also because he had an uncle who was an army colonel and could have pulled some strings. But Luria insisted on going into the military. He was not motivated by patriotism—by that time he was a committed anti-Fascist—but simply did not want any special treatment.

Although some of Luria's female friends thought he looked good in his uniform, army life was tedious and, at times, comical. When Luria took an oath of allegiance to the king of Italy in a colonel's office, he had to pull a saber out from its scabbard. Upon finishing the oath, he couldn't get the saber back in the scabbard and was awkwardly dismissed. On another occasion, Luria was supposed to join a battalion on a training exercise and was offered a horse. But he didn't know how to ride a horse, so instead, he hopped on a bicycle as the soldiers cheered.

Luria's experience as an army medical officer confirmed the fact that he did not want to practice medicine. With help from Ugo, he came up with a plan for the immediate future: he would go to Rome, finish his radiology courses and study physics. Ugo cleared the idea with physicist Enrico Fermi, who was working at the University of Rome. So in the fall of 1937, Luria hung up his army uniform, moved in with his aunt and uncle and started classes in Rome.

Luria's year with the physicists changed his life. Although he struggled with the physics classes because of the mathematics, he learned to think more analytically. Working with the no-nonsense Roman physicists also instilled in him a strong work ethic. But most importantly, through a series of articles by the German physicist Max Delbrück, he was introduced to the subjects of genetics and radiation biology. This was the link between biology and physics he had been looking for.

Now that Luria had a subject to dedicate himself to, exactly how should he proceed? At this point, a fortuitous incident occurred in the winter of 1938. Luria later called it the "trolley-car accident." He was riding the tram to work when it jerked to a stop due to an electrical failure. He recognized a fellow passenger from previous commutes and engaged him in conversation. The

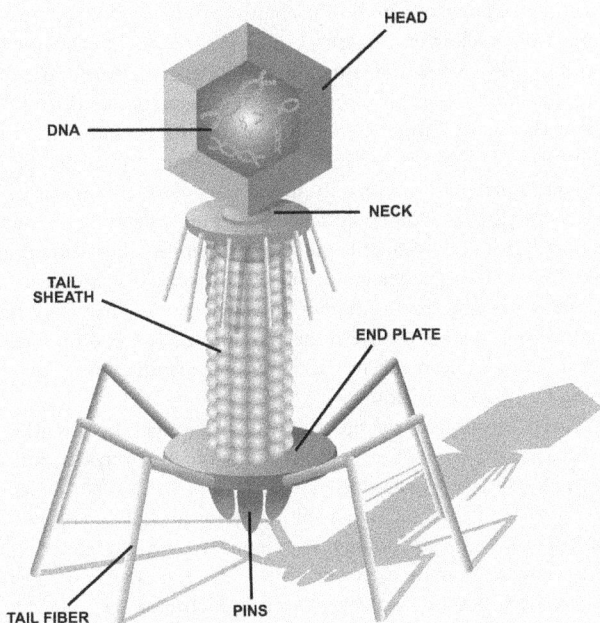

Structure of a bacteriophage. *Shutterstock.*

passenger turned out to be a bacteriologist named Geo Rita, and the talk turned to bacteria, genes and radiation. When the tram started again and Rita reached his destination, Luria accompanied him to his lab. There, Rita explained that he was testing water from the Tiber River for dysentery bacilli by looking for the presence of something called a bacteriophage ("phage," for short), a virus that infects and reproduces inside bacteria. Luria had never even heard of a bacteriophage, spider-like parasites so small they can only be seen using an electron microscope. They turned out to be perfect for testing Delbrück's ideas about genes. Bacteriophages were similar to genes, but easier to work with and one could make billions of copies in a few hours. Luria learned that Delbrück had escaped Nazi Germany, immigrated to the United States and was living in California working with bacteriophages at Caltech. As

Luria put it, "As in a troubadour romance, the love triangle among Delbrück, bacteriophage and me had come into being at four thousand miles' distance and without mutual awareness of each other."

Luria won a fellowship to travel to California and work with Delbrück. But in July 1938, Mussolini issued his "Manifesto of Race," which stripped Jews of Italian citizenship along with any governmental or professional positions they held. Luria's father lost most of his business, and his brother lost his job in industry. The family didn't have the means or the energy to immigrate. Luria had a choice to make: he could stay with his parents and return to medicine or he could leave Italy and pursue his dream of being a scientist. His family thought that Luria would be safer abroad and might be able to help them, so he chose to leave. Besides, it was easier for a single man, with no family responsibilities, to escape. So one November day, Luria packed his bag, said good-bye to his parents and hopped on a train bound for Paris. (Luria's family survived the war after spending two years during the German occupation in hiding.)

In Paris, Luria had the good fortune of meeting physicist Fernand Holweck, an expert in the fields of high-vacuum physics and radiation biology. Holweck was delighted that Luria, a physician, knew some physics, and within a week had secured a fellowship for his Italian visitor. Luria went to work in Holweck's lab, doing research on bacteria and bacteriophages. He lived in a couple of small rooms near the Curie Laboratory where he worked and spent his free time discovering French poetry and art.

Luria was exploring Brittany by bicycle when World War II broke out in September 1939. As the German army rolled toward Paris in May 1940, Luria remembered the heartbreaking spectacle of the evacuation: "The absence of chaos or real pain made it even more dismal—recalling one of those animal migrations that start accidentally and spread through the entire herd." On May 12, Luria and a friend walked around the Sacré-Coeur, marveling at the empty streets as they heard exploding artillery shells in the distance. Luria had saved enough money to travel to America and began his long journey that evening. He left Paris on a bicycle and, for a couple of days, stayed just ahead of the German troops. On two occasions, he was targeted by strafing planes.

His month-long journey took him first to the port city of Marseilles, an important stop because it had an American embassy. It took a month or two for Luria to obtain an American immigration visa, an exit visa from France and transit visas from Spain and Portugal. Luria recalled that in Marseilles, it was refugees versus bureaucrats. For Luria, "the battle of the visa was just one more experience, exercising my talent for survival." For some, the bureaucratic nightmare proved too much—they committed suicide on the embassy porches.

With his paperwork in hand, Luria traveled through Spain and Portugal by train to Lisbon, where he bought a ticket on a Greek ship sailing for New York. He left Europe on the SS *Nea Hellas* with fifty-two dollars and a new suit—and he was better off than most. In the early morning of September 12, 1940, Luria sailed into New York Harbor. "No city, before or after, has ever conveyed to me a similar sense of power," he later reminisced. "From the first day I felt I had come home."

Soon after his arrival, Luria visited Enrico Fermi at Columbia University. Fermi and his Jewish wife had also escaped Italy by going to Stockholm to pick up his Nobel Prize and then sailing to New York. Fermi had written Luria a letter of recommendation to the Rockefeller Foundation that resulted in a mix of mini-fellowships, and in just a few weeks, Luria found himself ensconced at the College of Physicians and Surgeons at Columbia. He was given a desk, a lab bench and an incubator on the fifteenth floor, where, from the window, he enjoyed a spectacular view of the Hudson River and the George Washington Bridge.

When he went to register as a candidate for citizenship, Luria was told he could change his name. He had been given the name Salvatore Edoardo Luria at birth but never liked his first name, so he split it into Salvador E., a Spanish name. When the immigration officer asked him what the *E* stood for, Luria turned and asked the next person in line for a name beginning with *E*. He said Edward. And so it was that Salvatore Edoardo Luria became Salvador Edward Luria.

In late December 1940, Luria finally met Max Delbrück in Philadelphia at a meeting of the American Physical Society, where they talked and made plans for the future. It was the beginning of a very productive professional relationship and personal friendship and also the beginning of what would become known as the "phage group." The pair spent the first of many summers at the Biological Laboratory at Cold Spring Harbor on the north shore of Long Island. In June 1941, the world's greatest geneticists and biochemists gathered there for a symposium. Afterward, the scientists spent the summer doing experiments, holding daily seminars, swimming and square-dancing. Luria was reunited with his old friend Ugo Fano who had also immigrated to the United States and was now a biophysicist working on the effects of radiation on genes. At the lab, Luria and Delbrück did their first experiments on how bacteriophage multiply within bacteria.

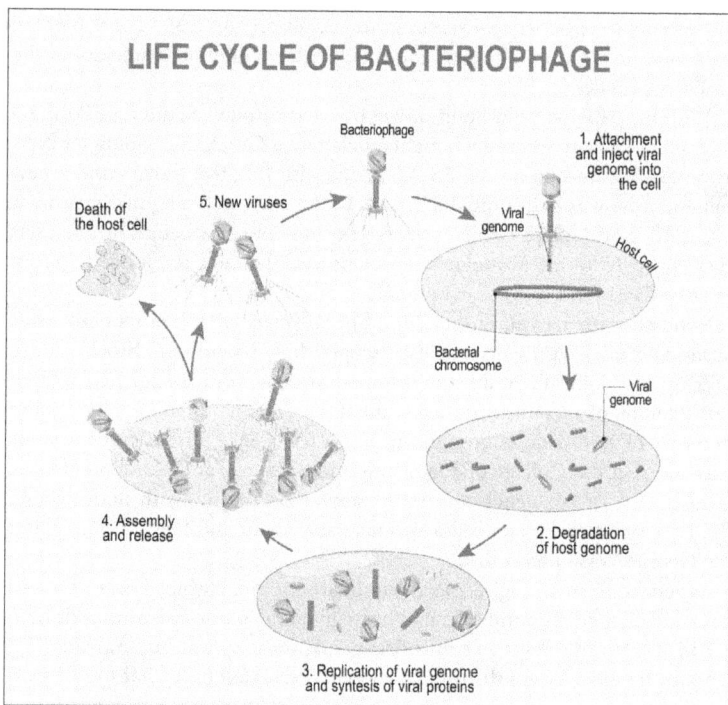

Life cycle of bacteriophage. *Shutterstock.*

Luria spent a short time working with Delbrück at Vanderbilt University. Then in the summer of 1942, Indiana University offered him an instructorship, and Luria moved to Bloomington. Luria claimed that Bloomington was the smallest town he had lived in for more than a few days and described it as "an ugly town, but not an uninteresting one." At the time, Indiana University was a leading center for the study of genetics with a distinguished faculty that included Ralph Cleland and Tracy Sonneborn. Nobel laureate Hermann Muller would join the faculty in 1945. On January 2, 1943, Luria began teaching bacteriology and eventually added virology and bacterial physiology to his repertoire. Luria expected to spend a semester in Bloomington and then be recruited by the army or asked to work on the war effort in some capacity. But the call never came, and Luria spent the next seven years at IU.

It was shortly after his arrival in Bloomington that Luria did the most important research of his life, work that would eventually lead to the Nobel Prize. The inspiration for Luria's experiment happened at, of all places, a faculty dance held at the Bloomington country club in February 1943. He found himself standing next to a slot machine and watching as a man put dimes into it. Although the man lost most of the time, occasionally the machine would spew some dimes back at him. Luria was teasing the man about his losses when suddenly he hit the jackpot, retrieved about three dollars in dimes, gave Luria a dirty look and walked away. It occurred to Luria that slot machine returns and bacterial mutations had something in common.

Excited, he immediately left the dance and early the next morning, a Sunday, hurried to his lab. He prepared several cultures of bacteria and then went to the library to anxiously wait for the cultures to grow. On Monday morning, he introduced the phage to the cultures and then busied himself with teaching. Tuesday was his "day of triumph." The results clearly showed that the phage-resistant bacteria came about through spontaneous mutations. He wrote to Delbrück, who responded by saying: "I believe you have something important. I am working out the mathematical theory." Days later, Delbrück had derived the mathematics that explained the results. The work became known as the Luria-Delbrück experiment, also known as the fluctuation test. It proved that Darwin's theory of natural selection based on random mutations applied to bacteria as well as higher forms of life. Of this time, Luria later said, "I was walking on clouds, just from the pleasure of the discovery." A second major discovery by Luria during his time at IU was the repair of radiation damage by bacteriophage. This opened up the field of DNA studies.

Luria's very first graduate student in 1947 was James Watson, who would go on to win the Nobel Prize for his discovery of the double-helical structure of DNA. Luria supervised Watson's doctoral research and was instrumental in arranging Watson's move from Copenhagen to Cambridge, a move that enabled Watson to pursue the research that led to his Nobel Prize. Another eventual Nobel laureate, Renato Dulbecco, came from Italy to work with Luria in 1947. Dulbecco stayed with Luria for two years as a research assistant before moving on to Caltech. Also in 1947, Luria became a naturalized citizen of the United States.

During his years in Bloomington, Luria met his wife, Zella Hurwitz. The meeting wasn't an accident. Zella's psychology professor at Brooklyn College knew she was going to IU to do graduate work, and as an acquaintance of Luria's, he also knew the geneticist was there. He suggested to Zella that she look him up—and she did. During their courtship, they had a couple of humorous misunderstandings. A few weeks after they met, Luria called and told her that he had two tickets to a football game and asked if she would like to go. She naturally assumed she would be going with Luria and said yes. Luria then explained that she would be going with one of his colleagues—he disliked sports and had no intention of attending. Zella changed her mind but didn't hold it against Luria. Later, Zella complained to a mutual friend that Luria, whom she assumed was an Italian Catholic, was telling Jewish jokes to her, a Jewish woman from the Bronx. When the friend told Luria, who was Jewish, they all shared a good laugh.

Luria and Zella were married in 1945 at the Monroe County Courthouse. Zella was twenty-one, twelve years younger than Luria. It was a good match—the couple had similar attitudes and opinions. Luria described Zella as "a woman with a remarkable mixture of strength, integrity, and reasonableness" who, after nearly forty years of marriage, "hardly ever told a lie to me or anyone else." They shared in the household responsibilities: one month, Zella would do the cooking, cleaning and chores, and the next month, Luria would do it. Luria admits there have been "occasional disagreement and grudges," but those were "soon resolved in open discussion and mutual trust." Zella earned a doctorate in psychology from IU in 1951 and joined Luria in academia. The couple had one child, a son named Daniel.

In his mid-thirties, Luria battled recurring bouts of depression, which he and his family suffered through. As he later recalled: "Depressive episodes of various intensities and durations came at intervals of several months for three decades, demanding of me a staggering effort, while they lasted, to keep functioning at a more or less normal pace." Although the depression did not significantly affect his work, the episodes made him tense and extremely critical and damaged his relationship with his son. Luria underwent years of psychotherapy, a treatment he later denounced as "intellectually vacuous." Luria complained that in the 1960s, when antidepressant drugs became available, his doctors did not prescribe them or prescribed them in much too small doses to be effective. Finally, under the care of a competent psychiatrist, he was given adequate doses of the antidepressant, which, together with lithium treatments, cured him of the depression.

In 1950, Luria got a job offer from the University of Illinois. The IU dean called to inform Luria that the offer would not be matched. Why? Because at the time, Luria was president of the University Teacher's Union, an organization that fought for improved salaries and working conditions. Luria gave what he thought was a mild speech supporting the union, but the local newspaper portrayed it as a full-throated denunciation of the university administration. That didn't sit well with the IU leadership, so Luria bid farewell to old IU and moved on to Illinois.

As the incident suggests, Luria was on the far left of the political spectrum, describing himself as a democratic socialist. According to Luria, his "commitment to socialism is essentially a commitment to the idea of economic justice and equality, a commitment fed by anger at the gross inequalities visible in society." Throughout his career, he was an outspoken supporter of left-wing political causes. In 1957, he was one of the ten initial signatories of a statement by chemist Linus Pauling against the testing of nuclear weapons. After the statement was released, in a clear attempt at intimidation, an agent from the Immigration and Naturalization Service showed up at Luria's house and started asking questions. In the 1960s, he opposed the Vietnam War and was briefly blacklisted from receiving funding from the National Institutes of Health. In the 1970s, he was involved in the debates about genetic engineering, where he advocated a middle-of-the-road approach with moderate regulations and oversight. And in the 1980s, he criticized the activities of the Israeli government, especially their involvement in Lebanon and Guatemala.

On the religious spectrum, Luria was a nonbeliever. He started questioning his Jewish faith when he was about twelve years old, gradually evolving into a deist and then, around the age of twenty, becoming a determined atheist. After witnessing the involvement of the liberal wing of the Christian Church in the civil rights movement during the 1960s, Luria softened his attitude toward religion, recognizing that it could be a force for social good. He also understood that religion was a medium for aesthetic experience through music, poetry and art. But he warned: "The bigotry of organized religion associated with unjust governments; the pernicious attempts to substitute the letter of so-called sacred texts for the observations of science; the use of such texts to justify inequality and oppression—these are the crimes of irrationality into which organized religion is constantly in danger of falling." His increased tolerance toward religious values made him all the more intolerant of the connection between organized religion and political power, an association that he claimed was "incompatible both with the sane exercise of power and with the legitimate claims of religion as an expression of an ideal of human brotherhood."

Luria characterized his nine years at Illinois as "fruitful." He had a big laboratory, worked with brilliant scientists and students and continued his research in phage genetics. In 1952, Luria discovered a phenomenon called the restriction modification, his third major discovery. He found that phages growing in an infected bacterium could be modified such that after they were released and reinfected a related bacterium the phage's growth is restricted or inhibited. The discovery happened because Luria accidentally broke a test tube containing the bacteria he intended to use, but rather than give up the experiment, used a completely different bacteria he thought would work. Luria jokingly called the incident his "lucky break." This phenomenon became the foundation of genetic engineering.

In 1959, Luria made a final move, this time to MIT, where he became chairman of the microbiology department. Zella became a professor at Tufts University, where she was an authority on the psychology of human sexuality and gender identity. At MIT, Luria's research focused on cell membranes and bacteriocins. (Bacteriocins are proteins produced by certain bacteria that inhibit the growth of closely related bacteria.) In 1963, while on sabbatical at the Pasteur Institute in Paris, Luria discovered that bacteriocins impair the function of cell membranes. Later, his lab at MIT found that this impairment happens because the bacteriocins form holes in the cell membrane, allowing ions to flow through and create havoc with the cell's electrochemistry.

Salvador Luria at his desk at MIT, circa 1969. *Wikimedia Commons.*

In 1969, Luria was awarded the Nobel Prize in Medicine or Physiology for "the replication mechanism and the genetic structure of viruses." He shared the prize with collaborators Max Delbrück and Alfred Hershey. Luria got the news while he was washing the breakfast dishes. Luria joked that he wished the Nobel committee had waited a bit longer because the prize money doubled the next year. At the ceremony in Stockholm, the banquet was announced by the sound of trumpets. Luria told Zella that it would be hard for him to sit down to eat from then on without the sound of trumpets. At his first dinner after returning home, Luria found a little plastic trumpet in his plate.

Luria became the director of MIT's Center for Cancer Research (CCR) in 1972. The CCR has produced four Nobel Prize winners, including Luria's colleague David Baltimore, who won the prize for his work on tumor viruses. In his later years, Luria's creative efforts turned toward writing. His book *Life: The Unfinished Experiment*, a popular science book on molecular biology, won the National Book Award for Science in 1974. Luria retired from MIT in 1978, although he remained active.

Salvador Luria died of a heart attack on February 6, 1991, at his home in Lexington, Massachusetts. He was seventy-eight years old.

EDWARD PURCELL

1912–1997

Field | Physics
Major Contribution | Won the Nobel Prize for Physics in 1952 for discovering nuclear magnetic resonance
Indiana Connection | Earned a degree in electrical engineering from Purdue University in 1933

> *I have not yet lost a feeling of wonder, and of delight, that this delicate motion should reside in all the things around us, revealing itself only to him who looks for it. I remember, in the winter of our first experiments, just seven years ago, looking on snow with new eyes. There the snow lay around my doorstep—great heaps of protons quietly precessing in the Earth's magnetic field. To see the world for a moment as something rich and strange is the private reward of many a discovery.*
> *—from Purcell's Nobel lecture, December 11, 1952*

EDWARD MILLS PURCELL was born on August 30, 1912, in Taylorville, Illinois, a small farming town in the central part of the state. According to a biographical memoir by Robert Pound, Purcell's father, Edward A. Purcell, grew up as a farm boy but became the manager of the local telephone company. He had a limited education but nevertheless spent some time teaching in a one-room country schoolhouse. In contrast, his mother, Mary Elizabeth Mills, was highly educated. She had a master's degree in classical studies from Vassar College north of New York City and was a high school Latin teacher. A younger brother joined the family five years after little Edward made his debut.

The home was filled with books that Purcell read as a child, and he spent considerable time at the public library. The family attended a Presbyterian church, where Purcell went to Sunday School, sang in the boys choir and "heard more sermons than I've heard in total since." He also recalled exploring the basement of the telephone office and finding sections of telephone cable and wire. He took the items home, melted the lead sheathing and removed the paper insulation from the wire, which he then used to make things. When he was older, he read the *Bell System Technical Journal* and, although he didn't understand the mathematics, admired the meticulously prepared articles adorned with beautiful diagrams. "It was a glimpse into some kind of wonderful world where electricity and mathematics and engineering and nice diagrams all came together," he later recalled.

When he was about fifteen, the family moved sixty miles east to Mattoon, where Purcell made friends with Dunlap McNair, who was fascinated by chemistry. The duo performed extracurricular chemistry experiments together, an activity that stimulated Purcell's interest in science and engineering. McNair had a father and an uncle who were very religious. The boys got into some heated theological debates with the men, which, as Purcell later recalled, "left us disgusted with the whole proposition." Although he might still be on the rolls of the Presbyterian Church, "I really have not considered myself an active church member since my bitter conflicts, intellectual conflicts, with Uncle Irving."

In his final two years of high school, Purcell took chemistry and physics, classes that "were rather important to me or seem so in retrospect." His chemistry teacher, Mr. Thomkins, "really knew some chemistry" and was the first adult Purcell had encountered who "was a real scientist." His physics teacher, Miss Edwards, did not know a lot of physics, but "she was a person who understood her own limitations and respected the subject and was utterly honest and sincere."

As an example of his physics teacher's forthrightness, Purcell tells the story of a physics problem they were assigned in which a man pulls himself up a flagpole. The man is seated on one end of a rope that runs over a pulley attached to the top of the flagpole and back down. The man hoists himself by pulling on the opposite end of the rope. The question is: how hard does the man have to pull? Purcell and McNair considered the problem and came up with an answer of half his weight. Miss Edwards, however, argued that the correct answer was his whole weight. But the boys didn't accept their teacher's answer and decided to put the question to a test. McNair had a large spring scale for weighing ice, and Purcell had a barn. They went into the barn and rigged up an apparatus similar to the one described in the problem. Purcell, who at the time weighed 120 pounds, sat down in the seat and started to pull on the opposite rope. When the scale read 60 pounds, he started to go up. Then McNair tried it and got similar results. The boys ran back to the school, found their teacher still in her room grading papers and announced their result. Her response was among the best in the annals of pedagogy: "Well, I must have been wrong and you must be right because you did an experiment and proved it." Miss Edwards may not have completely understood physics, but she understood science. "I've always felt that she did me a real service at that moment," Purcell later reminisced.

Purcell graduated high school in 1929, a few months before the onset of the Great Depression. Instead of choosing the University of Illinois only forty miles away, he opted for Purdue. As his major, Purcell, who had grown up surrounded with wires and telephone equipment, chose electrical engineering. He knew physics was a subject but had no idea that someone could actually be a physicist.

Purcell became aware of physics during his later years at Purdue. His first course in physics was given by a professor who, Purcell recalled, was "a lecturer of incomparable dullness." As a result, Purcell remembered almost nothing about the class. But in his junior year, Purcell noticed a course in the catalogue called "Independent Laboratory Work." No one had ever signed up for the course before—Purcell was the first. He was supervised by a professor named Walerstein who assigned Purcell the task of learning to use an old diffraction grating that had been stored in the attic of the physics building. After that, Purcell made an electrometer to measure the half-life of a radioactive substance. "I was really hooked at that point," Purcell later recalled.

During his senior year, Purcell worked under a graduate student named Hubert J. Yearian on the topic of electron diffraction, a phenomenon that had only been discovered a few years before. The work resulted in Purcell's first two scientific papers. The apparatus was in the basement of the physics building, and Purcell remembered that students were actually living down there, sleeping on cots and shaving in the sinks. It was the Great Depression, and some of the graduate students simply had nowhere else to go.

Purcell graduated from Purdue with a degree in electrical engineering in 1933 and decided to pursue an advanced degree in physics. One of his physics professors recommended him for a scholarship to study abroad, so Purcell spent the next year at the Technische Hochschule in Karlsruhe, Germany, where he studied atomic spectra. It was not a good time to be in Germany; Hitler had come to power just a few months before and things were beginning to change for the worse. One positive event that took place during Purcell's time in Europe was that he met his future wife, Beth Busser, a graduate of Bryn Mawr College who was on an exchange scholarship to study German literature. The couple met on the ship going over, and he visited her a couple of times in Germany.

The next year, Purcell accepted a full tuition scholarship at Harvard. During his first year, Purcell recalled taking a mathematics course called complex variable theory. Although he struggled in the class, he later claimed it was one of the great intellectual experiences of his life. In particular, he recalled taking the first semester final exam. He studied hard for the test and spent the night before proving every theorem in the course. But when he took the actual exam, he couldn't do anything—he froze—an experience many college students have had. He earned a C- for the first semester, a failing grade for a graduate course. But luckily at the time, grades were based on the entire academic year rather than a single semester. Purcell pulled his grade up during the second semester and finished the class with a respectable mark.

When the time came for his thesis work, Purcell focused on the problem of charged particles in a spherical condenser. His thesis advisor was Kenneth Bainbridge, an expert in mass spectroscopy who would later go on to direct the atomic bomb test. Purcell completed his research and earned his doctorate in 1938. Meanwhile, in 1937, he married Beth Busser. The couple would go on to have two sons, Dennis and Frank, born in the early 1940s.

Purcell stayed on at Harvard for two years as an instructor, teaching classes and working on a team building a cyclotron, a machine that smashes subatomic particles together. During World War II, Purcell was recruited to work down the street at MIT's Radiation Laboratory, known as the Rad Lab, where he spent the next five years. The Rad Lab's mission was to develop radar, a critical wartime tool. After the war, he stayed on at the Rad Lab to help write some of the books that summarized the wartime research.

His experiences at the Rad Lab greatly influenced Purcell's future career in physics. He later said that he acquired "a whole new kit of research tools" and had benefitted from "being thrown together in a working relationship with a number of physicists." The Rad Lab, along with the Manhattan Project, also marked the beginning of a new partnership between science and the military. Purcell put it this way: "You see, the scientific people evolved during World War II entirely new relationships with military people, one of mutual confidence and understanding of the problems and working together." After the war, Purcell became a member of the Air Force Science Advisory Board. Later, he served on science advisory committees to Presidents Eisenhower, Kennedy and Johnson. Regarding Purcell's service on the advisory committees, a fellow member recalled that he "did not speak often, but when he did there would be enormous silence in the room, because everybody knew whatever he said was going to be worth listening to with careful attention."

As the physicists at the Rad Lab contemplated a return to basic physics research, they began discussing what kinds of problems they would tackle. Purcell claimed that "it was out of that kind of talk that I was struck with the idea for what turned into nuclear magnetic resonance." Purcell, working with colleagues Robert Pound and Henry Torrey on evenings and weekends, discovered nuclear magnetic resonance (NMR) on Saturday afternoon, December 15, 1945. It was for this discovery that Purcell won the 1952 Nobel Prize for Physics. He shared the prize with Felix Bloch of Stanford University, who made the same discovery at roughly the same time independently of Purcell.

NMR is based on the fact that protons in the nucleus of an atom behave like tiny rotating magnets. When atoms or molecules are placed in a magnetic field, they align themselves with the field. Radio waves can disturb the direction of the alignment, but only in certain amounts determined by the laws of quantum mechanics. When the atoms return to their original positions, radio waves are emitted by the atoms with frequencies that are characteristic of different elements and isotopes. Thus, NMR can be used to determine the chemical composition, structure and properties of materials. NMR is also the basis of magnetic resonance imaging (MRI), one of the most important advances in the field of medicine in the twentieth century.

A few years later, Purcell made another discovery that is at least as important as the discovery of NMR. Dutch astronomers had predicted that hydrogen atoms in interstellar space would, by radiation or collision, be moved to a slightly higher energy level. When the atoms dropped back down to the lowest energy level, a radio wave would be emitted with a wavelength of twenty-one centimeters. But the Dutch scientists doubted that the wavelength would be detectable. Purcell and graduate student Harold "Doc" Ewen decided to give detection a try. Working in the evenings and on weekends, with borrowed equipment and a budget of $500, they built a

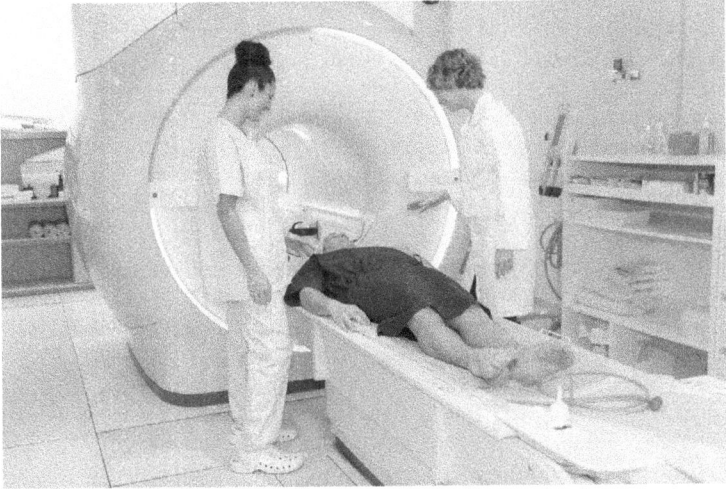

Patient undergoing an MRI scan. *Shutterstock.*

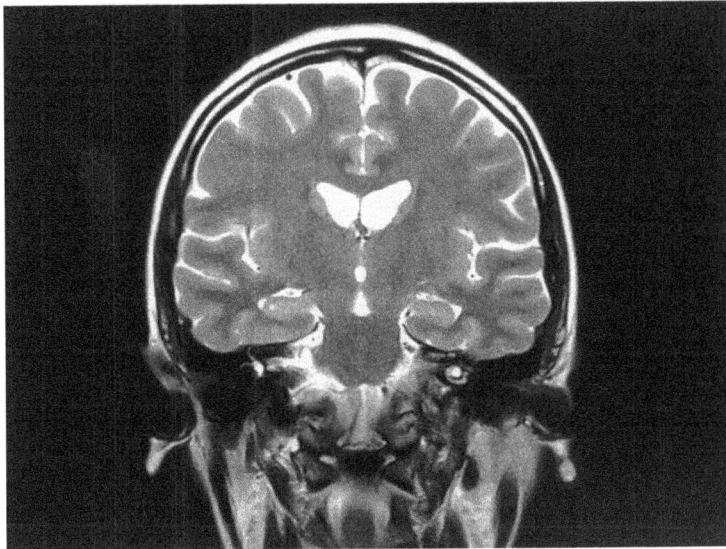

Magnetic Resonance Imaging (MRI) of a human brain. *Shutterstock.*

horn-shaped antenna and installed it outside a window on the fourth floor of Harvard's Lyman Laboratory. During heavy rains, the horn antenna funneled water into the building, and in the winter, the antenna was used as target practice for snowballs thrown by mischievous students. On March 25, 1951, they detected the wavelength. "It was just barely there," Purcell later recalled. The discovery was a major breakthrough in the field of radio astronomy and gave astronomers a way to map the large-scale distribution of matter in the galaxy.

In his later years, Purcell made noteworthy contributions to the field of biophysics. In 1977, working with Howard Berg, he did research on the motion of flagellated bacteria. Purcell proposed a successful explanation of bacterial locomotion involving helical flagella that rotate like corkscrews. One outgrowth of his work on bacterial motion was a famous lecture he gave in 1977 titled "Life at Low Reynolds Number." The Reynolds Number is a dimensionless quantity in fluid mechanics that helps predict the pattern of fluid flow. At low Reynolds number, fluid flow is mainly laminar or sheet-like. At high Reynolds number, fluid flow is marked by

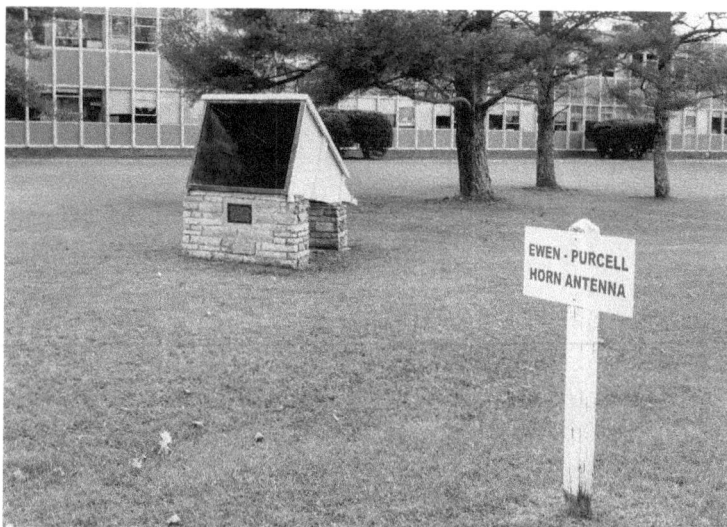

Ewen-Purcell Horn Antenna at the Green Bank Observatory in West Virginia. *Flickr Commons.*

turbulence. In the lecture, Purcell explained the forces and effects in fluid flow at low Reynolds number.

In addition to being a top-ranking scientist, Purcell also excelled in the classroom, where he exhibited a knack for explaining complex phenomena in simple ways. He wrote a textbook on electricity and magnetism that was notable for its use of relativity in the presentation. Of the many honors he received, he was especially proud of his Oersted Medal awarded by the American Association of Physics Teachers in recognition of his contributions to teaching.

Politically, Purcell was against wars and in favor of controlling nuclear weapons. He protested against the Vietnam War by resigning from the president's Science Advisory Committee in 1965 and ending his involvement on all military advisory councils. In September 1967, Purcell was persuaded to serve, albeit reluctantly, as spokesman for a group of Harvard faculty members who had been invited to meet with President Johnson at the White House. The meeting resulted from a letter the group had written outlining their reasons for opposing the war. Johnson was unconvinced by Purcell's arguments and proceeded to give a rambling, hour-long defense of the war. A committee member seated next to Johnson saw him scribble on a pad of paper "improvement relations Harvard University."

The quiet and modest Edward Purcell retired from teaching at Harvard in 1980, although he continued doing research for several years. In 1996, he fell and fractured his leg. During his recovery, he suffered from bacterial lung infections requiring long hospitalizations. He died of respiratory failure in his home in Cambridge, Massachusetts, on March 7, 1997. He was eighty-four.

13

BEN MOTTELSON

1926-PRESENT

Field | Physics

Major Contribution | Won the Nobel Prize for Physics in 1975 for his work on the structure of the nucleus.

Indiana Connection | Earned his bachelor of science degree from Purdue University in 1947.

> *We feel that in this cooperation we have been able to exploit possibilities that lie in a dialogue between kindred spirits that have been attuned through a long period of common experience and jointly developed understanding.*
> —*Mottelson in his Nobel Prize autobiography speaking about his collaboration with Aage Bohr*

BEN ROY MOTTELSON was born in Chicago on July 9, 1926, to Goodman Mottelson, an engineer, and Georgia Blum. He was the second of the couple's three children. Mottelson remembered his childhood home as "a place where scientific, political and moral issues were freely and vigorously discussed." He attended public schools in La Grange, a Chicago suburb, graduating from Lyons Township High School in 1943. World War II was raging, so Mottelson joined the U.S. Navy and was sent to the V-12 Navy College Training Program at Purdue University. The V-12 program was designed to produce officers for the navy by having students supplement their normal college classes with military courses. The war ended before he saw active duty, but he stayed at Purdue and earned his degree in 1947.

Mottelson pursued graduate studies in physics at Harvard University, where he worked under Julian Schwinger, an eventual Nobel laureate. While at Harvard, Mottelson married Nancy Jane Reno. The couple would go on to have three children, two sons and a daughter. After earning his doctorate in nuclear physics in 1950, Mottelson was awarded a traveling fellowship and used it to spend a year at the Institute for Theoretical Physics (now the Niels Bohr Institute) in Copenhagen, Denmark. At the time, the institute was run by the great theoretical physicist and Nobel laureate Niels Bohr and was one of the leading centers for the study of physics in the world. The following year, Mottelson was awarded another fellowship, this time from the U.S. Atomic Energy Commission, which allowed him to extend his stay for two more years. During his time in Denmark, Mottelson began a long collaboration with Aage Bohr, Niels Bohr's son, regarding distortions in the shape of the atomic nucleus.

At the time, there were two competing theories about the atomic nucleus. In the shell model, the protons and neutrons were arranged in concentric shells. In the liquid drop model, the particles behaved like a fluid. But neither one could explain all the known properties of the nucleus. In 1950, the American physicist James Rainwater postulated that the nucleus behaved more like a balloon filled with moving balls. Just as the balls could create distortions on the surface of the balloon, the moving protons and neutrons could create distortions on the surface of the nucleus. At about the same time, Aage Bohr reached a similar conclusion. Beginning in 1951, Bohr and Mottelson worked together to compare the new theoretical model with experimental data. In three scientific papers published from 1952 to 1953, Bohr and Mottelson showed that the theory agreed with experiment.

Also in 1953, Mottelson was appointed to the staff of European Organization for Nuclear Research (CERN) working with its Theoretical Study Group, which was conveniently based in Copenhagen. He stayed with CERN until 1957, when he joined a new organization, the Nordic Institute for Theoretical Physics (Nordita), as a professor, a position he would hold until his retirement. Bohr and Mottelson continued their collaboration by working on a two-volume monograph titled *Nuclear Structure*. The first volume, titled *Single-Particle Motion*, was published in 1969, and the second volume, titled *Nuclear Structure*, came out in 1975. In the meantime, Mottelson became a Danish citizen while maintaining his American citizenship.

Mottelson, Bohr and Rainwater shared the Nobel Prize for Physics in 1975 "for the discovery of the connection between collective motion and particle motion in atomic nuclei and the development of the theory of the structure of the atomic nucleus based on this connection." Niels Bohr and his son, Aage Bohr, thus became one of four father-son pairs to win a Nobel Prize in physics. Tragically, Mottelson's wife, Nancy, did not live to see her husband win the prize. She died a few months earlier from cancer.

In 1983, Mottelson married Britta Marger Siegumfeld. He continues to live in Denmark. In his free time, he enjoys bike riding, swimming and listening to music.

14

JAMES WATSON

1928-PRESENT

Field | Biology
Major Contribution | Won the Nobel Prize for Physiology or Medicine in 1962 for discovering the double helix structure of DNA
Indiana Connection | Earned his doctorate from Indiana University in 1950

> *One could not be a successful scientist without realizing that, in contrast to the popular conception supported by newspapers and mothers of scientists, a goodly number of scientists are not only narrow-minded and dull, but also just stupid.*
> *—James Watson in "The Double Helix," 18*

JAMES DEWEY WATSON was born in Chicago on April 6, 1928, shortly before the onset of the Great Depression, into a family "that believed in books, birds and the Democratic Party." He was the oldest child of James Watson Sr. and Jean Mitchell Watson; he was shy, she was social. Initially, the family lived in Chicago's Hyde Park neighborhood. Two years later, he was joined by a sister, Betty, and soon thereafter the family moved to South Shore, a middle-class community filled with bungalows and two-story apartment houses. After renting an apartment for several years, the family bought a small house at 7922 Luella Avenue where he and his sister slept in bunk-like beds in a little attic room. (The house, built in 1922, still stands.) Watson often could be found in an alley behind the house playing kick-the-can or setting off firecrackers. When he grew to be five feet tall, a basketball hoop was mounted on the garage; to pass the time in the winter, a ping-pong table was purchased, perhaps serving as a gentle introduction to tennis, a game that Watson enjoyed throughout his life.

Watson's father made a modest income as a debt collector for the La Salle Extension University, a thriving correspondence school offering business courses. His main task was to write letters dunning students for delinquent payments. His letters never made threats but instead simply reminded students of how their education would eventually lead to a better job. Watson's mother worked part time at the University of Chicago's housing office, helping to supplement his father's meager income, which had been cut in half by the Depression.

Both of Watson's parents were politically liberal. His mother was active in the Democratic Party, eventually becoming a precinct captain. The basement of their home served as the local polling station. The family fervently supported Franklin Roosevelt's New Deal and stood against the isolationists who sought to keep America out of the trouble in Europe.

Neither of Watson's parents was very religious. His father was an atheist, and Watson later claimed that "the luckiest thing that ever happened to me was that my father didn't believe in God." His mother was Catholic but only went to mass on Christmas Eve and Easter. For a while, Watson went to church with his grandmother, who lived with the family. He memorized the catechism and confessed his sins. But when Watson was ten, his father informed him that the Catholic Church supported the fascists in the Spanish Civil War, and that was the end of religion for Watson. After his confirmation at age eleven, he stopped attending mass and later described himself as "an escapee from the Catholic religion."

Instead of going to mass on Sunday mornings, Watson and his father went on birdwatching walks. Watson Sr. had been a devoted birdwatcher since high school and often ventured out on early morning walks in Chicago's Jackson Park. Interestingly, on one of these walks in 1919, Watson Sr. met a young college student who was equally fascinated by birds. The student's name was Nathan Leopold. They became friends and, in 1923, took a trip to Michigan in search of a rare bird called the Kirkland warbler. On the trip, they were accompanied by Leopold's friend Richard Loeb. Late in 1923, Leopold's father wrote to Watson Sr. expressing his concern about his son's strange obsession with Loeb. In May 1924, Leopold and Loeb, for a thrill, brutally murdered a young boy named Bobby Franks and dumped his body in a ditch. The pair were defended by Clarence Darrow, who asked Watson

Sr. to appear as a character witness. He declined. After they were convicted and sentenced, Leopold wrote to Watson Sr. from prison to propose a correspondence. He never replied to the letter.

The Leopold incident did not deter the Watsons from birdwatching. The first father-and-son bird walks were also in Chicago's Jackson Park, where Watson learned to distinguish between the various species of warblers, ducks, vireos and flycatchers. Because of his early fascination with birds, Watson set his sights on a career in ornithology.

Watson attended prekindergarten at the University of Chicago's Laboratory School, elementary school at the Horace Mann Grammar School and high school at South Shore High School, where he earned mostly top grades. At age fourteen, he appeared on *Quiz Kids*, a Chicago-based radio program featuring intellectually talented teenagers. He took the entrance exam for the University of Chicago and won a scholarship that paid for tuition.

In the summer of 1943, at the age of fifteen, Watson enrolled at the University of Chicago. He continued to live at home and took a half-hour ride on a streetcar to get to campus. Watson majored in zoology, earning mostly As and Bs, although he struggled with mathematics and got a C in calculus. His grades were good enough to get him elected to Phi Beta Kappa, an honor that thrilled his parents. Watson later wrote, "Never once did they [his parents] even hint that I was shirking my responsibilities in preferring books and birds over making money."

The single most important event during Watson's college days took place outside the classroom. One Sunday during his senior year, he read an article in the newspaper about a new book titled *What Is Life?* written by the Nobel Prize–winning physicist Erwin Schrödinger. The next morning, he ran to the library, checked out the book and began to read. The book explained how genes were the key to life because they carried hereditary information down through the generations. It made a profound impression on Watson, who later wrote, "As birds had bound me to life sciences, Schrodinger's exaltation of the gene would lead me to a life of studying genetics."

Watson earned his bachelor's degree in June 1947 and applied to three graduate schools: Caltech, Harvard and, following a suggestion from his advisor, Indiana. Caltech turned him down, perhaps because of his troubles with math. Harvard accepted him but offered no financial aid. Indiana University offered a fellowship and a $900 stipend on the condition that he would not study ornithology, a field for which they offered no graduate work. That was fine with Watson, who had shifted his focus to genetics. Besides, at the time, Indiana University was the equal of Caltech in the field of genetics and better than Harvard because no one there was even interested in the subject. So in September 1947, James Watson hopped aboard the Monon Railroad bound for Bloomington.

At Indiana University, Watson studied under biologist Hermann Muller, who had just won the Nobel Prize for discovering that X-rays can cause genetic mutations. His thesis advisor was Salvador Luria, who was studying the genetics of a simple type of organism called a bacteriophage ("phage" for short)—a form of a virus that multiplies inside bacteria. For Watson and his future research, it was a perfect group of scientists to work with. "I was trained to find the structure of DNA as Prince Charles is trained to be a king," Watson later said of his graduate work.

Watson spent his first summer as a graduate student at the Cold Spring Harbor Laboratory on Long Island, a facility he would later direct. There, he met Max Delbrück, a pioneer in the field of bacterial genetics and upon whose research the book *What Is Life?* was based. The second summer was spent at Caltech in Pasadena working with Delbrück. During that time, he also met Linus Pauling, the greatest chemist of the twentieth century and a future rival.

Watson received his doctorate from IU in May 1950 and spent the summer at Caltech and Cold Spring Harbor before boarding a ship to Copenhagen to begin a two-year National Research Council postdoctoral fellowship. In Denmark, Watson was supposed to work under Herman Kalckar, who wanted Watson to focus his research on the enzymes that create the precursors of DNA (deoxyribonucleic acid). Watson didn't think the research would lead anywhere, so instead, he spent his time at the State Serum Institute doing experiments on phages. Kalckar, preoccupied by marital problems, didn't seem to care. Watson finished his phage experiments late that winter and wrote up a paper. When Kalckar invited Watson to join him at the Zoological Station in Naples in April and May to learn about marine biology, he eagerly accepted.

During his time in Italy, Watson attended a lecture by an English physicist named Maurice Wilkins, a professor at King's College in London. Until then, Watson had assumed that no one, at least in the near future, would understand the three-dimensional atomic structure of DNA. The reason was that genetic information, encoded in the DNA, was variable. This implied that each DNA molecule would have a different structure. But at the end of his talk, Wilkins showed a photograph, taken using a technique called X-ray diffraction, suggesting that DNA had a regular, crystalline

Cold Spring Harbor. The laboratory campus is the cluster of buildings at the top of the hill. *Flickr Commons.*

structure. "Suddenly I was excited about chemistry," Watson later recalled. "Before Maurice's talk I had worried about the possibility that the gene might be fantastically irregular. Now, however, I knew that genes could crystallize; hence they must have a regular structure that could be solved in a straightforward fashion."

To uncover the structure of DNA, Watson had to first learn how to interpret X-ray diffraction images. Where could he go to learn how to do that? He tried to approach Wilkins but was given a cold shoulder. Caltech was out because it was doubtful that the great Linus Pauling would waste his valuable time on a mathematically challenged biologist. Then Watson remembered that he knew someone at Cambridge University interested in the topic and arranged to have his fellowship transferred to Cambridge's Cavendish Laboratory.

The twenty-three-year-old Watson arrived in Cambridge in the fall of 1951 to work with a small group of scientists doing research on the three-dimensional structure of proteins. Almost immediately, he met Francis Crick. The pair struck up a friendship and began sharing an office. Crick was thirty-five and, according to Watson, talked too much and laughed too loud. Yet he had a mind that could quickly penetrate to the essence

The Cavendish Laboratory at Cambridge University where Watson and Crick discovered the structure of DNA. *Shutterstock.*

of a problem. Like Watson, he had read Schrödinger's *What Is Life?* and because of it switched fields from physics to biology. Crick was also a friend of Maurice Wilkins in London. Over the next two years, Watson and Crick would work out the structure of the DNA molecule.

At the time, the DNA molecule was known to consist of a sugar molecule, a phosphate molecule and four bases: adenine, cytosine, guanine and thymine. But it was not known how these molecules all fit together. Using X-ray diffraction photographs as a guide, Watson and Crick built cardboard and metal models of the DNA molecule. But they weren't the only ones working on the problem. At Caltech, Linus Pauling was constructing his own DNA models, unaware that two young men in Cambridge were trying to beat him to the discovery.

Late in 1951, Watson and Crick thought they had the solution and called a meeting with their experimental collaborators Maurice Wilkins and Rosalind Franklin, an expert in X-ray crystallography. Wilkins and Franklin did not get along with each other. Why? Because the director of the lab had taken the DNA project away from Wilkins, who wasn't getting anywhere, and given it to Franklin. Franklin had been hired with the understanding that she would be studying the structure of DNA. The problem was that nobody told Wilkins. And nobody told Franklin that Wilkins was already studying DNA. When Wilkins began treating Franklin as if she was his assistant, Franklin wouldn't put up with it. And Wilkins just didn't know how to handle a strong woman.

Nevertheless, they tolerated each other's presence long enough to make it through the meeting. Things didn't go well—Wilkins and Franklin shot holes in the model. After the debacle, Watson and Crick were ordered to stop work on DNA—and they did, sort of. Crick went to work on his doctoral dissertation while Watson studied theoretical chemistry and viruses. But DNA remained a topic of conversation between the two, and Watson could often be found in the library leafing through scientific journals in search of a hint.

In early February 1953, a yet-to-be-published paper written by Pauling made its way across the Atlantic and into Watson's hands. It proposed a triple-helical structure for DNA. Almost immediately, Watson knew it was wrong and hurried off to London to inform Wilkins and Franklin about Pauling's failure. Wilkins was busy so Watson walked down the hall to Franklin's lab. The door was ajar, and Watson walked in. Franklin was startled and gave Watson a dirty look indicating that he should have knocked. Watson proceeded to tell her about Pauling's mistake. Watson kept

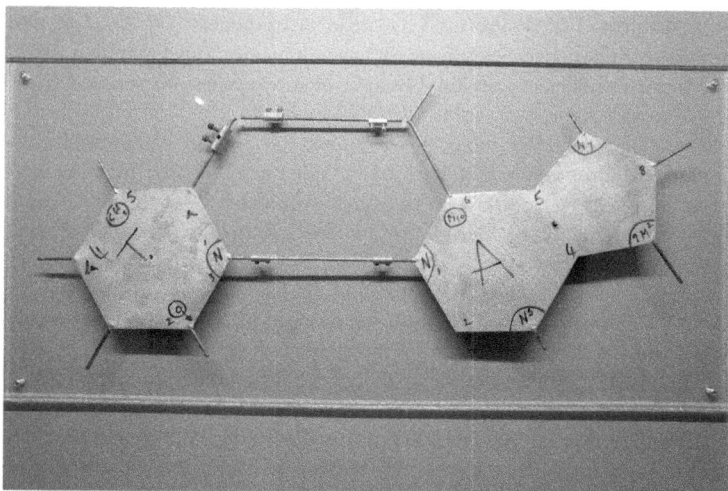

A section of Watson and Crick's original metal plate DNA model at the National Science Museum of London. *Shutterstock.*

mentioning helixes, Franklin kept objecting and the discussion got heated. When Watson implied that she was an incompetent interpreter of X-ray diffraction photographs, Franklin made an aggressive move toward him. Afraid that she was going to hit him, Watson moved toward the door, only to find Wilkins standing there, innocently unaware of what was going on. The two men exited the room, and Franklin slammed the door. As they walked down the hall, Watson told Wilkins what had happened.

The awkward incident prompted Wilkins to open up to Watson. He opened a desk drawer and took out a photograph that had been given to him by Franklin's assistant. It was an X-ray crystallography photo (now known as "Photo 51") that she had taken back in the summer. Franklin had no idea her data was being shared with Watson. "The instant I saw the picture," Watson later wrote, "my mouth fell open and my pulse began to race. The pattern was unbelievably simpler than those previously obtained."

Why was the image simpler? Franklin had previously discovered that DNA came in two forms: A and B. The A form was drier, more crystalline, and showed greater detail in photographs. The B form was wetter and produced a simpler image. Watson was now looking at a photograph of the simpler B form that showed very clearly a helical structure. Most of the previous

X-ray photographs had been mixtures of the two forms, which made any interpretation more difficult.

This was the breakthrough Watson and Crick needed. Armed with the new information, Watson and Crick threw themselves at the problem. It took less than a month to put the pieces together. While working with a cardboard model on his desk early on Saturday, February 28, Watson had a final key insight: "Suddenly I became aware that an adenine-thymine pair held together by two hydrogen bonds was identical in shape to a guanine-cytosine pair held together by at least two hydrogen bonds." He put the finishing touches to the model and showed it to Crick. Later that day, Crick went to The Eagle pub and triumphantly announced that he and Watson had "discovered the secret of life."

They ran it past Wilkins and Franklin, who agreed the structure was correct. When Linus Pauling heard about the discovery, he was thrilled. Watson and Crick wrote a brief nine-hundred-word paper titled "A Structure for Deoxyribose Nucleic Acid," which was published in the scientific journal *Nature* on April 25, 1953, along with articles by Wilkins and Franklin. A

The Eagle Pub in Cambridge, England, where the discovery of the structure of DNA was first announced. *Andy Oxford/Wikimedia Commons.*

A historical plaque at the Eagle Pub. *Chris Sampson/Wikimedia Commons.*

longer, more detailed paper appeared on May 30. It was one of the most important scientific discoveries of the twentieth century, but the world at large hardly took notice. Elizabeth II was about to be crowned the queen of England, and Sir Edmund Hillary was about to climb Mount Everest. But over time, the supreme significance of the discovery began to soak in.

The structure Watson and Crick discovered is known as the double helix—sort of a twisted ladder. The sides of the ladder are made from alternating sugar and phosphate molecules. The rungs of the ladder are formed by the four bases bonded together by hydrogen. The bases always bond together in the same way: adenine with thymine and guanine with cytosine. The genetic information is encoded by the sequencing of the bases. By breaking the bonds between the bases—that is, by "unzipping" the molecule through the middle—the molecule can make a copy of itself. In a breathtaking example of scientific understatement, Watson and Crick noted this feature in their first short paper: "It has not escaped our notice that the specific pairing we have postulated immediately suggests a possible copying mechanism for the genetic material."

James Watson, along with Francis Crick and Maurice Wilkins, won the 1962 Nobel Prize for Medicine or Physiology. The Nobel citation reads, "For their discoveries concerning the molecular structure of nucleic acids and its significance for information transfer in living material." What about Rosalind Franklin? Tragically, Franklin died in 1958 of ovarian cancer at the age of thirty-seven, and Nobel Prizes are not awarded posthumously. The cancer was probably caused by her exposure to the X-rays she used in her work. She rarely wore a lead apron and often walked right through the beam. Without Franklin, Watson and Crick would have gotten nowhere.

After the discovery, Crick continued to work at Cambridge and became a driving force behind advances in molecular biology. Watson went to Caltech to work with Max Delbrück and others. In 1956, Watson accepted a position at Harvard University, where he was a professor for twenty years. His research at Harvard focused on RNA (ribonucleic acid) and the role it played in the transfer of genetic information. He championed a shift away from the traditional biology that concerned itself with plants and animals and toward a biology that focused on molecules and cells using the language of chemistry and physics.

James Watson and Francis Crick on the right shaking hands with geneticist Maclyn McCarty. McCarty was born in South Bend. *Marjorie McCarty/Wikimedia Commons.*

Watson was not well liked by some of his colleagues at Harvard. One of the distinguished professors that Watson irritated the most was Edward Osbourne Wilson, usually known as E.O. Wilson. At the time, both were young men in their late twenties; Watson was already famous, whereas Wilson's fame would come later. Wilson once called Watson "the most unpleasant human being I had ever met." According to Wilson, Watson thought that traditional biology "was infested by stamp collectors who lacked the wit to transform their subject into a modern science." At biology department meetings, Watson "radiated contempt in all directions." But people put up with Watson's bad manners because of the greatness of his discovery.

After winning the Nobel Prize in 1962, Watson published very little scientific research and turned his attention to writing. He penned two important biology textbooks: *The Molecular Biology of the Gene*, published in 1965, and *The Molecular Biology of the Cell*, published in 1983. In 1968, he published *The Double Helix*, a popular and personal account of the events leading up to the discovery of the structure of DNA. It revealed how science is really done—it showed the messy side of science. Doing science involved luck, ego, dead ends, wrong answers and clashing personalities. The book is listed by the Board of the Modern Library as number seven on its list of the one hundred best nonfiction books of the twentieth century and was named one of eighty-eight "Books that Shaped America" by the Library of Congress.

The book was a best-seller, but it was also controversial. In fact, it was originally supposed to be published by the Harvard University Press, but when the book was sent to Francis Crick and Maurice Wilkins to be read and reviewed, they strongly objected to the way Watson portrayed them. The press took the highly unusual step of canceling publication, so Watson published the book commercially. The tone of the book is evident in its very first sentence: "I have never seen Francis Crick in a modest mood." Crick wouldn't speak to Watson for several years after the book was published.

The book paints an especially harsh portrait of Rosalind Franklin, who comes off as secretive and difficult. But as Watson explains in the preface, the book is an attempt to re-create his first impressions of the relevant events and personalities and does not take into account what he had learned since. The book is an autobiography, not a history. In the epilogue, he directly addresses the Rosalind Franklin issue, saying: "Since my initial impressions of her, both scientific and personal...were often wrong, I want to say something here about her achievements." He goes on to list her accomplishments and says that he, Crick and Franklin later became friends. Watson also writes that he and Crick realized "years too late the struggles that the intelligent woman faces to be accepted by a scientific world which often regards women as mere diversions from serious thinking."

Watson always had a keen interest in the opposite sex, an interest that was not always reciprocated. After his great discovery, Watson became obsessed with finding a wife. In 1968, after fifteen years of romantic disappointments, he finally found the love of his life. Her name was Elizabeth "Liz" Lewis, a nineteen-year-old college student who had been helping out in his office. They were married on March 28, 1968, in La Jolla, California.

Also in 1968, Watson was named director of the Cold Spring Harbor Laboratory (CSHL), a place he had loved ever since his first summer visit during graduate school. For years, he commuted back and forth between Cambridge and CSHL, but by 1974, the Watsons had made Cold Spring Harbor their permanent residence. By 1976, the energetic Watson was no longer able to juggle a Harvard professorship, directorship of CSHL and a family, so he resigned from Harvard. Under Watson's leadership, CSHL expanded its research focus and created science education programs. He transformed the small facility into one of the world's leading research and education institutions. He also helped raise funds for the laboratory and would often untie his shoes and muss up his hair to polish his "mad scientist" persona. Today, CSHL has about 1,100 employees and students with a research budget of $140 million. Areas of research include cancer, neuroscience, genomics and quantitative biology and plant biology.

By the early 1970s, the Watsons had two sons: Rufus Robert Watson in 1970 and Duncan James Watson in 1972. By the time Rufus turned three, the couple had begun to notice that he had trouble interacting with people. By age ten, he was having social problems at school and a child psychologist was consulted. In the tenth grade, Rufus had a full-blown psychotic episode at school and was sent home. A few days later, Rufus went to the top of the World Trade Center in New York City, intending to end his life. After the incident, he was diagnosed with schizophrenia, and James Watson did something his wife had never seen him do before or since—he cried. Liz wanted to make sure Rufus felt loved, but Watson wanted something more—he wanted Rufus cured.

It was Watson's quest to understand the genetic basis for mental illness that provided the personal impetus for a 1986 symposium at CSHL on the question of mapping the human genome. Watson had called together the world's leading geneticists to discuss the issue, but the night before the meeting, Rufus ran off. Watson missed part of the meeting because he and Liz were out looking for Rufus. Nevertheless, that meeting and others like it eventually led to the Human Genome Project, an international effort to map the base pairs that make up human DNA. And who better to lead the project than James Watson, who, at CSHL, had proved his skill as an administrator.

In 1990, Watson was appointed as head of the Human Genome Project at the National Institutes of Health (NIH). But Watson held that position only until April 1992, when a public conflict with the new NIH director and boss, Bernadine Healy, made it impossible for him to continue. Healy wanted to acquire patents of gene sequences, and Watson was violently against it, arguing that the human genome belonged to the people, not nations. Watson, who always said exactly what he thought, called Healy a lunatic, not once but several times, in public. That brought an end to Watson's association with the project. After Watson resigned, Frances Collins took over, and in 2000, the Human Genome Project was completed. In 2007, Watson became the second person ever to publish his fully sequenced genome online. He did it "to encourage the development of an era of personalized medicine, in which information contained in our genomes can be used to identify and prevent disease and to create individualized medical therapies."

Eventually, the issue regarding the patenting of DNA was settled by the courts. In 2013, the U.S. Supreme Court unanimously ruled that naturally occurring human genes are not patentable because DNA is a "product of nature." The court did allow patenting for DNA that had been manipulated in the lab because such DNA is not found in nature. Specifically, the court ruled that a type of DNA known as complementary DNA (cDNA) could be patented.

In his later years, Watson made several highly controversial public remarks concerning racial differences in intelligence in which he repeatedly asserted that differences in IQ scores between Black and white people are due to genetics. As a result, the CSHL called his remarks "unsubstantiated and reckless," revoked all of his honorary titles and cut all remaining ties with him, although he continues to live on the grounds.

15

EI-ICHI NEGISHI

1935-PRESENT

Field | Chemistry
Major Contribution | Won the Nobel Prize for Chemistry in 2010 for developing methods in organic synthesis.
Indiana Connection | Instructor at Purdue University from 1968 to 1972, professor from 1979 until his retirement

> *Each evening, I would study until after 11 pm when I heard my mother's gentle reminder saying "Isn't it about time to stop studying and go to bed?" In retrospect, I feel very fortunate that neither of my parents ever told me to study more or harder in my life. For one thing, they themselves were busy just for our survival, even though I did amply sense their silent but strong mental support for my higher education.*
> —*Ei-ichi Negishi in his Nobel Prize autobiography*

EI-ICHI NEGISHI was born in Changchun, in northern China on July 14, 1935. At the time, the region, known as Manchuria, was occupied by the Japanese and Negishi was a Japanese citizen. When he was one year old, the family moved to Harbin, where he lived for the next eight years. He began elementary school at age six, a year earlier than normal. When Negishi was nine, the family moved yet again, this time to Seoul, South Korea, also occupied by Japan. In November 1945, a few months after the end of World

War II, the family returned to Japan and moved into a house in Tokyo that had somehow survived the massive Allied bombings. Japan was devastated by the war, and food shortages were common. The large Negishi family, with five children ranging in age from one to twelve, had to find a way to feed themselves. So they moved a final time to a one-acre patch of land about thirty miles from the center of Tokyo where the family could grow the food they needed. It was here that Negishi spent his junior high, high school and college years.

Negishi was accepted into an elite high school focused on getting as many students as possible accepted into Japan's top-rated universities, including the prestigious University of Tokyo. Initially, Negishi thought he didn't stand a chance. After his first year, he ranked 123rd out of over 400 students at the school. He reasoned that while there were about 100 students ahead of him, there were 300 behind him. He concluded that if he studied as hard as he could, he might just have a shot at the University of Tokyo.

From that point forward, Negishi devoted every extra minute to studying. Each morning, he woke up a couple of hours early to prepare for his classes, and every evening, he studied late into the night. Negishi's hard work paid off. By the end of the first semester of the eleventh grade, his academic rank had jumped dramatically to ninth. By the end of the year, he was first and maintained that rank throughout the twelfth grade. When the time came to take the entrance exam, he was feeling intense pressure almost to the point of getting sick. After he took the two-day exam, he was more than halfway sure that he had failed. But he passed, and became, at age seventeen, one of the youngest college entrants into the rigid Japanese system. Although Japan's high-stakes exam system has often been criticized, Negishi concludes that it is highly effective training for professional careers, especially in science and engineering. He claims that even decades later, he often uses something he learned during his preparation for the exams.

Negishi's first two years at the Komaba campus of the University of Tokyo was a bit of a disappointment for him. The years were full of general education courses that included law, economics and psychology. According to Negishi, the students weren't very interested in learning, nor were the instructors very interested in teaching. For example, he had to take a foreign language and chose German. One German class was devoted to grammar, while in another simultaneous course, students were exposed to German poems and novels. Negishi remembered struggling to read and interpret Goethe while having to look up nearly every word in a Japanese–German dictionary—not a very effective way of learning a language. As a result, Negishi's study habits deteriorated.

During this time, Negishi pursued two musical extracurricular activities. The first was listening to Western classical music by the great composers. The second was conducting and singing in a small choir that met at the house of his junior high school music teacher, Tsuguo Suzuki. Suzuki had a daughter named Sumire, who caught the eye of Negishi. They started dating during his freshman year; eventually, they would marry.

Despite his lax study habits, at the end of his freshman year Negishi was in the top third of the 450 or so students pursuing a degree in physical science and engineering. This qualified him to choose one of most coveted majors: applied or industrial chemistry. At the time in the mid-1950s, the Japanese economy was booming, partly as a consequence of the war on the nearby Korean peninsula. In applied chemistry, the synthetic polymer industry was growing rapidly, attracting young scientists and engineers into the field.

Negishi's junior year turned out to be tough going for two reasons. First, it was a two-hour one-way commute between his home and the Hongo campus of the University of Tokyo where his classes met. Four hours a day standing on a crowded commuter train was exhausting. Secondly, his classes, lasting from eight in the morning until five in the afternoon, were poorly taught. The stress resulted in poor health, and he ended up in the hospital for several weeks. He missed his midyear exams and was forced to repeat his junior year.

Negishi now considers this lost time as a "blessing in disguise" because it gave him time for thinking, planning and reading. He read a wide range of literature from various "how to" books to the Bible, although he wasn't a Christian. This time of personal reflection led him to conclude that the road to happiness had four essential requirements, which he enumerated in an autobiographical sketch after winning the Nobel Prize. The requirements were: 1) good health; 2) happy surroundings, including one's own family and beyond; 3) selection and pursuit of a worthy professional career; and 4) one or more enjoyable and lasting hobbies.

Renewed and refreshed, Negishi restarted his junior year in April 1956. He avoided the long commute by renting a small room near the Hongo campus. He also learned conversational English, a skill he considered important for his career. Musical activities were pursued on the weekends, and he continued to date Sumire with the intention of marrying her. During the second half of his junior year, Negishi applied for and received a lucrative scholarship from Teijin Ltd., a company specializing in the manufacture of synthetic polymers. Under the terms of the scholarship, he would join the company after graduation. The scholarship eliminated financial worries during the last year of college. In March 1958, Negishi graduated from the University of Tokyo with a degree in chemical engineering. On the same day, at a little restaurant near campus where their families had gathered to celebrate, Negishi and Sumire announced their engagement.

Now a research chemist at Teijin's Iwakuni Research Laboratories near Hiroshima, Negishi was assigned the task of coming up with polymers that had particular properties. He soon discovered that his preparation was woefully inadequate and decided that he needed to improve his understanding of synthetic organic chemistry. He remembered the welcoming speech by the company's president mentioned a Fulbright scholarship that would pay for up to three years of study in the United States. The president pledged that if anyone should win the scholarship, the company would grant a leave of absence and provide additional financial assistance. Negishi decided to apply.

Applicants for the scholarship would be tested on their English proficiency, with a difficult two-part exam covering both written and conversational English. Negishi had already been studying English for several years, and the company hired a tutor to help him prepare. Out of 150 applicants who took the exam, Negishi was one of two chosen for the program. After an eight-week English orientation class in Hawaii in August and September 1960, Negishi set off to study at the University of Pennsylvania, the school chosen for him by the Fulbright Commission. Negishi said that the opportunity to study in the United States was "the single most important turning point in my professional career."

As a first-year graduate student, he earned eight consecutive excellent grades on the organic chemistry cumulative exams, a feat unheard of in those days. This boosted his confidence in his research ability. During his time at Penn, he heard lectures from about a dozen current and future Nobel Prize winners, including Glenn Seaborg, Linus Pauling and Herbert Brown, the latter of whom would become his mentor. It occurred to Negishi that if he kept working in the right direction, he just might someday win a Nobel Prize.

In December 1963, Negishi earned his doctorate degree and returned to Japan and the Teijin company. He had decided to become an academic researcher at a Japanese university but could not find a position. In 1966, he resigned from Teijin and accepted a postdoctoral appointment at Purdue University, working under future Nobel laureate Herbert Brown.

Ei-ichi Negishi at a Nobel Prize press conference in Stockholm in 2010. *Holger Motzkau Wikipedia/ Wikimedia Commons.*

In 1968, he became Brown's assistant with the rank of instructor. Negishi moved on to Syracuse University in 1972 but returned to Purdue in 1979 as a full professor. Soon after he arrived, his mentor, Herbert Brown, won the Nobel Prize. Brown was the first faculty member at Purdue to receive the award, but he was not the last.

Negishi won the 2010 Nobel Prize in chemistry for work he began in the mid-1970s. The prize was shared with the American chemist Richard F. Heck and fellow Japanese chemist Akira Suzuki. Negishi related the story of getting a phone call one morning and reaching for the phone while trying to think of who could be calling at that early hour. As he brought the phone to his ear, he suddenly thought, "Wait a minute—this is the day!" And then he heard a voice with a Scandinavian accent on the other end confirming the good news: "Good morning. You have been chosen as the winner of the Nobel Prize." That moment changed Negishi's life. Later that same morning, the whole campus was abuzz with excitement. Purdue is the only Big Ten school to have two Nobel Prizes in chemistry.

The Nobel citation states that the prize was awarded "for palladium catalyzed cross couplings in organic synthesis." To understand this discovery, we first have to consider what is meant by "organic synthesis." Organic chemistry is the chemistry of the carbon atom, the most versatile atom on the periodic table and the basis of life on earth. Synthesis refers to building molecules in the laboratory. So in organic synthesis, one is simply building carbon-based molecules in the lab. A catalyst is a substance that speeds up a chemical reaction without being consumed by the reaction. So putting it all together, Negishi developed chemical reactions in which carbon atoms, helped by palladium, are linked or coupled together. This allows for the easy and efficient synthesis of complex organic molecules and found applications in the pharmaceutical and electronics industries.

After his retirement, Negishi thought about returning to Japan but decided instead to stay at Purdue. Negishi found the right balance between science and family life, was always home for dinner with his two daughters and was always supported by his loving wife, Sumire. The Nobel Prize medal sits on a table at his home in West Lafayette.

Ei-ichi's life story took a tragic turn when, early in the morning of March 13, 2018, he was found by police on an Illinois state highway south of Rockford. The day before, the couple had been reported missing by family members. According to police, he was walking along the road in a confused state. A short time later, police found the body of Sumire in a landfill. An autopsy revealed that she had died of hyperthermia, although Parkinson's disease and hypertension were contributing factors. Evidently, the couple was trying to find the airport and got stuck in a ditch on a road near the landfill. Ei-ichi, in a state of confusion and shock, was trying to find help. No foul play was ever suspected.

16

FERID MURAD

1936-PRESENT

Field | Medicine
Major Accomplishment | Won the Nobel Prize for Physiology or
  Medicine in 1998 for discoveries concerning nitric oxide as a
  signaling molecule in the cardiovascular system
Indiana Connection | Born and raised in Whiting, graduated from
  DePauw University

> *I like to ask a question that's never been asked before,
> get the answer, and be the only person in the world with
> the answer to that question. And hopefully, if I'm lucky,
> it's an important question.*
> *—quote from Murad, accepting an honorary doctorate
> from DePauw*

FERID MURAD (later known as "Fred") was born in Whiting, a town in
northwest Indiana, on September 14, 1936. According to Murad's Nobel
Prize autobiography, his father, Jabir Murat Ejupi, an immigrant from
Albania, was born into a family of shepherds. Jabir had less than a year
of formal education but spoke seven languages. As a teenager, Jabir ran
away from home to sell candy in neighboring countries. He met a group of
teenagers in Austria, and together, they decided to immigrate to the United
States. He landed at Ellis Island in August 1913, where the immigration
officer asked what his name was. It was not unusual for names to be

Americanized and so it was with Jabir, who the officer declared was now John Murad.

Ferid's mother, Henrietta Josephine Bowman, was born in Alton, Illinois. She attended grade school for a few years before quitting to help raise her younger siblings while her mother went to work. In 1935, at age seventeen, Henrietta ran away from home to marry John, who was thirty-nine. The next year, Ferid was born, joined by brothers John in 1938 and Turhon in 1944. The three boys were raised in a four-room apartment behind their parent's restaurant in Whiting. As soon as he was old enough, Ferid helped out in the restaurant by washing dishes, waiting tables and running the cash register. He worked every night and on weekends throughout grade school and high school. He made a game of memorizing each customer's order, adding up the bill mentally and meeting them at the register with the total. As is typical with family-owned businesses, his parents worked long hours, usually putting in sixteen-to-eighteen-hour days. Ferid vowed to get a good education so he wouldn't have to work as hard as his parents.

Ferid had an easy time in school and didn't study much. Yet he was an honor student every semester and graduated fifth in his high school class. When he was assigned to write an essay about his three top career choices in the eighth grade, he chose: 1) physician, 2) teacher and 3) pharmacist. So Ferid knew, from an early age, what he wanted to do with his life. He had an opportunity to attend the University of Chicago after his junior year but decided instead to finish high school, a decision he later looked back on with no regrets.

With his high school diploma in hand, he considered his options for college. Since his parents couldn't afford tuition, he looked for a school that would offer substantial financial aid. DePauw University awarded Ferid a hefty scholarship and he accepted, making him the first in his family to go to college. He entered the school in 1954 as a double major in premed and chemistry. To help cover his expenses, Ferid waited tables, taught the anatomy and embryology labs and worked in the summer. Due to the distractions of being a fraternity pledge, his first-year grades were a mixed bag of As and Bs with one C. But with growing self-confidence and better study habits, his grades gradually improved. By his senior year, most of his grades were As, and he was elected to Phi Beta Kappa.

In 1957, Ferid met his future wife on a spring break trip to Fort Lauderdale. Her name was Carol Ann Leopold, a DePauw Spanish and English major from St. Louis. They began dating, although most of their outings were "study dates" at the library because that was all Ferid could afford. They were engaged at Christmas and got married in June 1958, a few weeks after graduating.

During his senior year, Ferid learned about a new program at Western Reserve University (now Case Western Reserve University) in Cleveland that allowed students to earn a medical degree and a doctorate at the same time. The program paid full tuition for both programs along with a modest $2,000 yearly stipend. Ferid was interviewed by the entire pharmacological department in February 1958 and accepted the offer. His new fiancée was a little concerned about his facing another seven years of school, but nevertheless, she supported his decision. The plan was for her to teach high school English as he went through the program, but their plan hit a snag when she became pregnant shortly after their marriage. She taught for only a semester before she was asked to resign because her pregnancy was showing. As the years went by, she worked as a substitute teacher, a part-time secretary and a hospital clinic coordinator. She also gave birth to four daughters, including a set of identical twins. A fifth child, a boy, was born in 1967.

To help support his large family, Ferid moonlighted at the Cleveland Clinic one or two nights each week. He worked in the OB-GYN department, where he performed pelvic exams, assisted with Caesarian sections and helped with deliveries. The twelve-hour shift ran from 7:00 p.m. to 7:00 a.m. and paid twenty dollars an hour. Sometimes, he would have a full day of classes after working all night. The busy schedule sometimes kept him away from his family for four or five nights each week. Ferid managed to have dinner with his brood when he could and, during the summer, in conjunction with scientific meetings, took them camping for several weeks all over the United States.

Ferid worked closely with Earl Sutherland Jr., head of the pharmacology department, and Theodore Rall, Sutherland's collaborator. According to Ferid, Sutherland was "a visionary who was able to bring together multiple disciplines and areas to apply to his work." Rall taught Ferid how to do large and complicated experiments with the proper controls that yielded publishable results. Within the group, the research efforts became known as "Sunday experiments." The university was trying out a new integrated organ-system approach to medical education that helped Ferid learn and absorb the vast amount of information covered in his classes. He also enjoyed his clinical rotations in pediatrics, OB-GYN, orthopedics, surgery and neurology. Ferid later recalled that he "was in my element and loved it." He was at the top of his class every year in both graduate school and medical school and, upon graduation in 1965, won awards for both clinical medicine and research.

With a doctorate in pharmacology and a medical degree, Ferid decided to do his residency and internship at Massachusetts General Hospital, one of the top hospitals in the country. There, Ferid worked with some of the best and brightest medical minds in the world and later claimed that he "couldn't have asked for a greater introduction to medicine." He did miss his laboratory work and often found himself in the library reading abstracts on the topic of second messengers and hormone signaling to keep up with developments in the field. Based on his reading, Ferid kept a notebook describing future experiments he wanted to do.

After completing his residency in 1967, Ferid took a position as a clinical associate at the National Institutes of Health (NIH), where he worked in the laboratory of biochemist Martha Vaughn. She gave Ferid the freedom to pursue his own areas of interest, and he was able to perform many of the experiments he had earlier envisioned. Another of Ferid's mentors at the NIH was Vaughn's husband, Jack Orloff, a renal specialist. Ferid described Orloff as "superficially a gruff and tough man" but a "sensitive person and talented scientist."

Ferid spent more than three years at the NIH before he was recruited by the University of Virginia to start a new clinical pharmacology division in the School of Medicine. He would also become an associate professor in medicine and pharmacology. At thirty-three years old with five children to support, Ferid decided it was time to get a real job and accepted the offer. He joined the faculty in September 1970 and launched his own independent research career focused on the biological effects of nitric oxide, work that would eventually win him the Nobel Prize.

In July 1981, Ferid left Virginia to take a job as chief of medicine at Palo Alto Veterans Hospital, a hospital affiliated with Stanford University. Despite administrative duties and a heavy teaching load, he still maintained a research laboratory that employed about fifteen students, staff and fellows from all over the world.

Ferid left Stanford in 1988 for the private sector and became vice president of Abbott Laboratories, a medical device and healthcare company. He went to Abbott primarily because of the company's president, Jack Schuler, a Stanford MBA who, according to Ferid, had "considerable vision." The two men worked well together: Schuler taught Ferid about business and Ferid taught Schuler about discovering and developing drugs. Ferid took advantage of Abbott's scientists, facilities and money. During his four-year tenure, Abbott introduced two dozen drugs for clinical trials in treating various diseases. Before Ferid came on board, Abbott had no postdoctoral fellows or outside funding; when he left, Abbott had thirty-five fellows doing pharmaceutical research and about $3.5 million in outside funding. Then Abbott decided to reorganize its upper management, and Ferid's business associates were shown the door. Now Ferid found himself caught between management, marketing and the scientists and was constantly having to defend his research decisions. Ferid later explained that there were "always considerable marketing pressures on me that in my opinion were often the wrong decisions to develop novel therapeutics for diseases without adequate therapy."

Faced with an untenable position at Abbott, Ferid left and decided to start his own company. In 1993, he became a founder, president and CEO of Molecular Geriatrics Corporation, a research-based biotechnology company. Unfortunately, the investment banker in charge of financing the company couldn't raise the money he promised and lost a fortune on the venture. Murad had to travel around the world seeking investors and partners so he could pay the bills and keep the company afloat. After partnering with a major pharmaceutical company, the chastened Murad left the private sector and returned to academia. In April 1997, he accepted a position with the University of Texas Medical School at Houston where he formed a new department of integrative biology, pharmacology and physiology.

The next year, in 1998, Murad won the Nobel Prize for Physiology or Medicine, a prize he shared with Robert F. Furchgott and Louis J. Ignarro, both professors at UCLA. According to the Nobel Prize citation, the prize was awarded to the trio "for their discoveries concerning nitric oxide as a signaling molecule in the cardiovascular system." The fact that nitroglycerin caused blood vessels to expand had been known since the days of Alfred Nobel himself. Murad helped figure out exactly how this worked. It turns out that nitroglycerin activates an enzyme that forms cyclic guanosine monophosphate (cGMP), which causes expansion of the blood vessels. This increases blood flow. Back in 1976, Murad showed that nitroglycerin formed cGMP by emitting nitric oxide, a gas also known as nitrogen monoxide (NO). (Nitric oxide should not be confused with nitrous oxide, an anesthetic commonly known as laughing gas.) Nitric oxide is highly reactive and diffuses quickly across cell membranes, creating a mechanism by which cells can communicate or "signal" with each other. A gas as a signal-transferring molecule had not been observed before. The inner lining of blood vessels (endothelium) uses the nitric oxide to signal the surrounding muscle to relax, thus allowing the vessels to expand or dilate, resulting in increased blood flow.

Murad's discovery has been applied in medicine to regulate blood pressure, treat heart and lung conditions and fight infections. But the best-known application is in the treatment of erectile disfunction (ED) in men. The drugs that treat ED don't actually produce nitric oxide but instead work by enhancing the signaling mechanism. Murad jokes that the media often

Ferid Murad with fellow Nobel laureate James Watson (*far right*) and Dr. Zhdanov (*middle*) at a 2010 award ceremony in Moscow. *Vera Knorre/Wikimedia Commons.*

wants to talk about Viagra, but he attempts to steer the discussion into more important areas, medically speaking.

After winning the Nobel Prize, Murad traveled the world lecturing and making appearances. He noted that upon becoming a Nobel laureate, you are considered an expert on everything, as evidenced by invitations to serve on panels discussing world peace and education. One of his more memorable experiences was serving as grand marshal of the Fourth of July parade in his hometown of Whiting. He also made several trips to Macedonia, his father's homeland. On one of these visits, he arranged for one of his daughters to adopt an Albanian baby girl.

In 2011, Murad moved to George Washington University in Washington, D.C., as a professor in the department of biochemistry and molecular biology. His duties included teaching a course for undergraduates, mentoring graduate and medical students and continuing his research. Upon accepting the appointment, Murad said, "I think I have something to offer young people that gets them excited about medicine. I love research. I love to answer tough questions. I love to figure out how this information can be beneficial in clinical medicine to treat people." In 2016, Murad moved to the Palo Alto Veterans Medical Center, a hospital affiliated with Stanford University, to teach and conduct research.

17

RICHARD SCHROCK

1945-PRESENT

Field | Chemistry
Major Contribution | Won the Nobel Prize for Chemistry in 2005 for discovering metal-based catalysts
Indiana Connection | Born in Berne, Indiana, and lived in Decatur until age fourteen

> *I still find the process of unlocking nature's secrets an enormously satisfying profession and hope to be fortunate enough to continue to practice it for some time.*
> *—from Schrock's Nobel Prize autobiography*

RICHARD ROYCE SCHROCK was born in Berne, in northeastern Indiana, on January 4, 1945. The town is named after the capital of Switzerland and has a replica of Bern's famous clocktower. Schrock's parents, Noah J. Schrock, a carpenter, and Martha A. Habegger, were married in 1933 during the Great Depression. Neither parent had, at the time, completed high school, but eventually went back and earned their diplomas, an act that evidenced the value they placed on education. Schrock has two older brothers: Luther, born in 1934, and Theodore, born in 1939. Luther would grow up to be an engineer while Theodore chose medicine and became a surgeon. A few months after Richard was born, the family moved to Decatur, Indiana, about thirteen miles north of Berne. Schrock recalled that the house was close to the city swimming pool, where he "spent many happy summer days."

When Schrock was five, the family moved into an old house on the west side of South Thirteenth Street in Decatur. His father put his carpentry skills to work and, over several years, renovated the home. The house was situated on a large one-acre lot, and Schrock remembered that the yard "took forever to mow" on hot summer days. The family kept a garden that yielded corn, tomatoes, melons, strawberries and raspberries. In the summer, Schrock explored the local woods and ponds, fished, caught frogs and snakes and built little huts from sticks and twine. In the winter, he learned how to ice skate. They never had much money, but the house was warm and the food was plentiful for the family of five.

Schrock was introduced to chemistry on his eighth birthday when his brother Theodore gave him a chemistry set as a present. The simple gift changed Richard's life. "I was hooked," Schrock recalled. "I created a small laboratory at the end of a storage area for canned goods and used my budding woodworking skills to build shelves for the ever expanding collection of test tubes, beakers, and flasks. I obtained most of my equipment through a mail order supply house with money earned from an early morning paper route." He did simple experiments—mixing acids and bases together to make salts and creating esters with a nice smell—by following the procedures in hand-me-down chemistry lab books. At first, he used a simple alcohol burner, then moved up to a common Bunsen burner, which was eventually replaced by a high-tech broad-headed model capable of producing enough heat to melt metal.

When the family moved again to a house on Jackson Street, Schrock's chemistry laboratory occupied a small basement room and grew in size and sophistication. The local chemistry teacher gave him some surplus equipment and textbooks to add to his collection. Although he now had access to some potentially dangerous chemicals, there were no major mishaps. Schrock's mother, however, remembered an incident where the fire department was called to the home to put out a small fire that had started on a rug.

Noah Schrock went to San Diego, California, in the fall of 1958 to work in the construction industry. The next year, the rest of the family joined him, ending their time in Indiana. Richard navigated while his mother drove the car with the little chemistry laboratory packed securely in the trunk. Schrock's interest in chemistry grew in San Diego. He bought equipment at a laboratory supply house and purchased chemicals at a drugstore. He did a project that investigated osmotic processes in sea urchin eggs and entered it in a regional science fair, where he won a prize. The eggs had been harvested from sea urchins that Schrock had collected at low tide.

Schrock graduated from Mission Bay High School in 1963. The only colleges the family could afford were public universities with in-state tuition, so he applied to the University of California. He was accepted at Berkeley but chose instead to attend the Riverside campus, east of Los Angeles, thinking that a smaller school might enable him to do more research as an undergraduate. He was right about that. After doing well on his first exam in his chemistry course, his professor recruited him for a summer job. The research was in the area of atmospheric chemistry, an important topic in the smog-prone region of Los Angeles. Most of his time was spent blowing glass, constructing vacuum lines and measuring the concentrations of gases. The work eventually resulted in getting his name on two scientific papers, the first of over four hundred.

As Schrock's years at Riverside came to an end in 1967, one of his professors told him that he had enough ability to attend Harvard University. He applied and was accepted. An interest in music had been stirred by a Western civilization class, so he celebrated by loudly playing Rachmaninoff's second piano concerto, which he listened to through giant homemade speakers. When he arrived at Harvard, he was surprised to learn that a childhood friend from Decatur, David Swickard, was there to study political science. The two old friends shared a room in Perkins Hall, the dormitory for graduate students. The next year, Schrock moved to an apartment near the Central Square in Cambridge, where he was joined by several more friends. The apartment was cheap, furnished and convenient to the lab where Schrock spent most of his time. He lived there for the next three years, at times uncertain about how many roommates he had.

Schrock knew he wanted to focus on physical chemistry but was unsure about a research topic. One day he talked to a recently arrived assistant professor named John Osborn, who told him all about transition metal chemistry and the new and colorful crystalline compounds that could be created. The idea of making potentially useful new compounds excited Schrock, and he signed on. Schrock chose to work with rhodium and, for his dissertation, discovered several new compounds.

One November evening in 1969, Schrock found himself at a party where he met a tall, blond, beautiful and bright schoolteacher named Nancy Carlson. The couple started dating, and in August 1971, after finishing their degrees (for her, a master's degree in library science), they were married. They would go on to have two sons and enjoyed a happy life together.

Because permanent jobs for chemists were rare in 1971, Schrock accepted a postdoctoral fellowship from the National Science Foundation to work at Cambridge University, where he did research in the laboratory of Jack

Lewis. It was the beginning of a long and fruitful relationship with English scientists. The next year, Schrock returned to the United States and landed a job with DuPont at its Central Research Department (CRD) in Wilmington, Delaware. The CRD was unusual in that it was an academic department in an industrial setting, an entity that doesn't exist today.

At DuPont, Schrock's group leader gave him the freedom to pursue his own interests and allowed him to work after regular hours. It was sometime during the fall of 1972 that Schrock first heard the term *olefin metathesis*, the topic that would eventually win him the Nobel Prize. Olefins (also called alkenes) are compounds made of hydrogen and carbon (hydrocarbons) that contain one or more pairs of carbon atoms linked by a double bond. Metathesis is an important type of chemical reaction involved in the synthesis of organic substances. He began to read the metathesis literature and suspected that some chemicals he discovered in 1973 might be relevant to the process. In late 1974, he invited a professor from MIT to give a talk at DuPont. After Schrock explained his own research to his distinguished visitor, he got a job offer from MIT. He couldn't resist an academic position at a leading university and jumped at the opportunity even though it came with a lower salary.

In August 1975, Schrock and his wife moved to Boston, where he joined the faculty at MIT. By the end of his third year, Schrock had a group of ten students working in his lab. His research initially focused on the element tantalum, a rare, hard, bluish-gray, corrosion-resistant, chemically inert transition metal. Its main use is in electronic components, mainly capacitors and resistors. By 1980, he had applied the principles underlying tantalum chemistry to the elements tungsten, molybdenum and rhenium. In the mid-1980s, Schrock did research in polymer and dinitrogen chemistry. In the mid-1990s, he and a colleague applied a certain type of metathesis reaction to organic chemistry.

In October 2005, Schrock got the call from Sweden that every scientist hopes for. He had won the Nobel Prize for Chemistry, sharing the prize with French chemist Yves Chauvin and American chemist Robert Grubbs. The prize was awarded "for the development of the metathesis method in organic synthesis." What does that mean? Recall that organic chemicals, the foundation for all life, are chemical compounds that contain the element carbon. Organic synthesis is the intentional building of organic compounds. As previously stated, metathesis is a kind of chemical reaction in synthesizing organic substances. More specifically, in metathesis, double bonds between carbon atoms are broken in a way that causes different groups of atoms

Richard Schrock giving a TED talk in 2014. *Christine Racz/Wikimedia Commons.*

within the molecules to change places with one another. This rearrangement yields new molecules with different properties.

Schrock tried to find a way to make the metathesis reaction go faster. In other words, he tried to find a catalyst, a substance added to a chemical reaction that increases the rate of the reaction without being consumed. Working with a mechanism that had first been proposed in the early 1970s, Schrock systematically tested catalysts containing tantalum, tungsten or other metals to see which ones worked. The big breakthrough came in 1990, when Schrock discovered an efficient metathesis catalyst that used the metal molybdenum. The chemistry was based on a class of metal-containing compounds called Schrock carbenes that he had been working on since the 1970s. Practically speaking, Schrock's catalysts help make the production of fuels, plastics and pharmaceuticals cleaner and more efficient: cleaner because there's less hazardous waste, more efficient because it cuts the number of required steps by two-thirds. Thus, metathesis catalysts played a role in what is now known as "green chemistry."

Schrock is a co-founder and member of the board of a Swiss-based company that develops and applies proprietary metathesis catalysts. As of 2006, he had published around 425 scientific papers and trained about sixty-five graduate students and seventy-five postdoctoral students.

18

ERIC WIESCHAUS

1947–PRESENT

Field | Molecular Biology
Major Contribution | Won the Nobel Prize for Physiology or Medicine in 1995 for his discovery of the genes that control embryonic development in fruit flies
Indiana Connection | Born in South Bend, graduated from the University of Notre Dame

> *We did it almost as a classical genetics experiment. It could have been done anytime from the 1930s onward but we did it at a time when other people were discovering how to clone genes. If we'd done it earlier, it would have been just as good an experiment but the intellectual impact would have been less. But because of cloning and molecular approaches that were suddenly possible in flies, here were all these interesting genes and their postulated functions just begging for molecular analysis. It attracted huge numbers of really bright people.*
> *—Wieschaus in a 2007 interview with the American Society for Cell Biology*

According to his Nobel Prize autobiography, **ERIC FRANCIS WIESCHAUS** was born in South Bend, Indiana, on June 8, 1947. When Eric was six, his father, a chemical engineer, was transferred to Birmingham, Alabama. Eric recalled that he and his four siblings (a brother and three sisters) would go exploring in the nearby woods, collecting frogs, crayfish and turtles from a creek. He attended Catholic schools and did well in his science and math classes but was more interested in drawing and painting and dreamed of someday becoming an artist.

That all changed during the summer before his senior year when he went to Lawrence, Kansas, to attend a program funded by the National Science Foundation to encourage young people to consider a career in science. (This was the post-Sputnik era, and the United States was trying to build an army of scientists and engineers to keep up with the Russians.) Wieschaus found himself surrounded by teens who were as smart or smarter than he was—a group that talked about books, art and science. He felt a deep sense of belonging and was able to overcome his natural shyness and insecurity. For the first time, in the zoology lab, he dissected animals from fish to fetal pigs and examined their internal organs. He was invited back the next summer to work in a neurobiology lab, where he removed nerves from tortoises, stripped off the protective sheath and submitted them to electrical stimulation. The experiences convinced Wieschaus to become a scientist and major in biology.

Wieschaus returned to South Bend to attend the University of Notre Dame. During his sophomore year, he got a job preparing fly food in Professor Harvey Bender's Drosophila (fruit fly) lab, where he learned basic genetics. Wieschaus also took an embryology class, where he witnessed the wonder of cell cleavage, the first several cell divisions of the new embryo. He watched the intricate rearrangement of cells in frog embryos during a process called gastrulation, where the embryo changes from a simple spherical ball of cells into a multilayered structure. His observations provoked questions. Why did the cells behave that way? What were the mechanisms that made the cells different from each other? What forces caused these cellular rearrangements?

During his last years at Notre Dame, Wieschaus became active in the antiwar movement. He participated in protests and collected signatures on petitions. He also applied for conscientious objector status to avoid military service, although he realized that he was unlikely to be granted the request since he had not been raised in a pacifist religion. When Wieschaus graduated from Notre Dame in 1969, Professor Bender intervened and wrote to Donald Poulson, a geneticist at Yale, explaining Wieschaus's draft predicament. Poulson took Wieschaus on as a graduate student sight unseen.

It was a tumultuous time in the country and on the Yale campus. Several members of the Black Panthers were on trial for kidnapping and murder, bombs went off in the campus skating rink and the United States bombed

Cambodia. But Wieschaus loved New Haven and reveled in the political turmoil. In the classroom, Wieschaus nearly failed a cell biology class and was in danger of being kicked out of graduate school altogether. It was, he later admitted, the low point in his career. He ended up with a C, and his professor continued to nod at him in the hall, a gesture he took to mean everything was okay.

During his first year at Yale, Wieschaus worked in Poulson's Drosophila lab, where he learned that flies had embryos and that they go through the same sort of changes that the embryos of other animals go through. In his second year at Yale, with Poulson nearing retirement, Wieschaus switched to Walter Gehring's lab. Gehring was starting a new fly lab, and Wieschaus was his only student. This gave Wieschaus the unique opportunity to work one-on-one with Gehring. He learned how to do experimental science and how to grow embryos in living organisms. Two years later, Gehring moved his lab to Basel, Switzerland, and Wieschaus followed him, continuing work on his doctoral degree, which he earned in 1974. Wieschaus decided to become a cultural European. He learned some French, spoke fluent German and read about the history of Switzerland.

Wieschaus met two women in Switzerland, both of whom would play important roles in his life. The first was Christiane "Janni" Nüsslein-Volhard, a German developmental biologist whom Wieschaus met in Basel. She shared Wieschaus's interest in Drosophila embryology and worked with him in the lab. "Eric was loved by the technicians," she recalled. "Every Sunday he brought a hot meal he had cooked to the lab, walking the fifteen minutes through the woods with his big bag. I usually brought in a cake. When we had dull repetitive work to do, we listened to 'The Magic Flute.'" After moving to Zurich to do postdoctoral work, Wieschaus often returned to Basel, partly to finish experiments, but also to meet with Janni and plan future research. "She was the single most important influence in my work," Wieschaus claimed. "And she is still a close friend."

In Zurich, Wieschaus began performing Drosophila experiments with a Swiss graduate student named Trudi Schüpbach. The pair worked on the genetics of sex determination in fruit flies, an effort that required working late nights in the lab. The collaboration evolved first into a close friendship and then into a romantic relationship. "It was proximity," Wieschaus later explained. "We started as colleagues." The two scientists would eventually marry.

In 1978, Wieschaus began work at the new European Molecular Biology Laboratory (EMBL) in Heidelberg, Germany, designed as an international meeting and working place for scientists throughout Europe. Wieschaus was named a group leader, a position that gave him the freedom to pursue his interest in embryos without being encumbered by teaching duties or unnecessary paperwork. The job was an extraordinary opportunity, and he regrets that more young scientists don't enjoy such an experience at the beginning of their careers. But the most important aspect of his new job in Heidelberg was that Nüsslein-Volhard had also been offered a position. Now they could actually carry out the experiments they had talked about back in Basel.

At EMBL, Wieschaus and Nüsslein-Volhard collaborated on the work that would eventually win them the Nobel Prize. The experiment involved a process called mutagenesis in which the genetic information in an organism is changed, resulting in a mutation. Specifically, they looked at genetic mutations that affected embryonic development. This enabled them to specify the genes that drove the process. In the course of doing the experiment, they also developed a particular type of genetic screen, now known as the Heidelberg screen. A genetic screen is an experimental technique used to identify and select individual organisms with specific observable characteristics. Looking back, Wieschaus claimed that his years in Heidelberg "were probably the most exciting, intellectually stimulating ones of my entire scientific career."

Wieschaus returned to the United States in 1981 to accept a position at Princeton University doing research and teaching genetics and embryology classes. Wieschaus was joined at Princeton by his colleague and romantic partner Trudi Schüpbach, who became a professor of molecular biology. The couple married in 1983. Wieschaus says that life with her and their three daughters has "provided a needed balance to the demands of the lab."

Early one October morning in 1995, Wieschaus received a long-distance phone call from Sweden informing him that he had been awarded the Nobel Prize. He insisted that they had the wrong number, but it was no mistake. Wieschaus shared the Nobel Prize for Medicine or Physiology with Christiane Nüsslein-Volhard, his colleague from Heidelberg, and Edward B. Lewis, an American geneticist. According to the Nobel Prize citation, the prize was awarded "for their discoveries concerning the genetic control of early embryonic development." To understand this, we begin with the fact when a fertilized egg divides, at first the cells are all identical. But over time, they start to change. Some cells become part of the brain, others part of the heart, and still others become part of the nervous system. This process is controlled by genes. By experimenting with fruit flies, Wieschaus and Nüsslein-Volhard identified the fifteen genes that determine the size, shape and position of a cell during embryo development.

Why is this important? When the results were published in 1984, the Drosophila community took notice, but no one else did. Then, in the mid-1980s, other researchers perfected the cloning of genes and the results of the Wieschaus–Nüsslein-Volhard experiments took on added significance. Eventually, the genes found in flies were found to have similar or identical matches in humans. Mutations in the genes discovered by Wieschaus can change the fruit fly's normal body plan. Similarly, mutations in the corresponding human genes can change the normal human body plan. This discovery changed how scientists look at congenital birth defects in humans.

In 2007, Wieschaus described his life as follows: "I still work in the lab. I don't give talks that much. I'm around Princeton a lot. I still teach. My life is pretty much like it was before. People here are willing to tolerate my choices because they say, 'Oh, gosh, he's the Nobel Laureate and if that's what he wants, well, we'll let him have it.' Besides, in my case, what I want is pretty cheap." Also in 2007, Wieschaus was one of seventy-seven Nobel laureates to sign a petition in favor of repealing the Louisiana Science Education Act, a controversial law that allows teachers to use classroom materials critical of the theory of evolution. As of this writing, Wieschaus continues in his role as a professor of molecular biology at Princeton.

BIBLIOGRAPHY

Chapter 1: Science in Indiana Begins in New Harmony

Warren, Leonard. *Maclure of New Harmony.* Bloomington: Indiana University Press, 2009.

Chapter 2: Harvey Wiley

Blum, Deborah. *The Poison Squad.* New York: Penguin Books, 2018.
Wiley, Harvey W. *Harvey W. Wiley: An Autobiography.* Brooklyn, NY: Bobbs-Merrill Company, 1930.

Chapter 3: Vesto Slipher

Hoyt, William Graves. *Vesto Melvin Slipher 1875–1969: A Biographical Memoir.* Washington, D.C.: National Academy of Sciences, 1980.

Chapter 4: Hermann Muller

Carlson, Elof Axel. "Hermann Joseph Muller." In *Biographical Memoirs*, 188–218. Washington, D.C.: National Academies Press, 2009.
Davidson, Keay. *Carl Sagan: A Life.* New York: John Wiley & Sons, 1999.

Chapter 5: Harold Urey

Arnold, James R., Jacob Bigeleisen and Clyde A. Hutchison Jr. "Harold Clayton Urey." In *Biographical Memoirs*, 363–411. Washington, D. C.: The National Academies Press, Volume 68, 1995.

Rhodes, Richard. *The Making of the Atomic Bomb*. New York: Simon and Shuster, 1986.

Shindell, Matthew. *The Life and Science of Harold C. Urey*. Chicago: University of Chicago Press, 2019.

Chapter 6: Alfred Kinsey

Gathorne-Hardy, Jonathon. *Kinsey: Sex, the Measure of All Things*. Bloomington: Indiana University Press, 1998.

Chapter 7: Percy Julian

Cobb, William Montague. "Percy L. Julian Memorial Lecture." DePauw University, April 28, 1977.

"Percy Julian: Forgotten Genius." *Nova*. Directed by Llewellyn M. Smith. 2007. PBS.

Taylor, David. "Percy Julian: A Scientist Makes Inroads in Chemistry and Civil Rights." *Humanities* 28, no. 1 (January/February 2007).

Chapter 8: Wendell Stanley

Colvig, Ray. "Wendell M. Stanley Obituary." *Cancer* 29, no. 2 (February 1972).

Chapter 9: Emil Konopinski

Emery, Guy, Lawrence Langer and Roger Newton. "Emil J. Konopinski Obituary." *Physics Today* 44, no. 10 (1990).

Rhodes, Richard. *The Making of the Atomic Bomb*. New York: Simon and Shuster, 1986.

Chapter 10: Herbert Brown

Negishi, Ei-ichi. *Herbert Charles Brown 1912–2004: A Biographical Memoir.* Washington, D.C.: National Academy of Sciences, 2008.

Nobel Prize. "Herbert Brown Biographical." https://www.nobelprize.org/prizes/chemistry/1979/brown/biographical.

Chapter 11: Salvador Luria

Luria, S.E. *A Slot Machine, A Broken Test Tube: An Autobiography.* New York: Harper & Row, Publishers, 1984.

Chapter 12: Edward Purcell

Pound, Robert V. *Edward Mills Purcell: A Biographical Memoir.* Washington, D.C.: National Academies Press, Biographical Memoirs, Volume 78, 2000.

Purcell, Edward. *New York Times* obituary, March 10, 1997.

Sopka, Katherine. Transcript of interview with Edward Purcell. Neils Bohr Library & Archives, Harvard University, November 23, 1976.

Chapter 13: Ben Mottelson

Nobel Prize. "Ben Mottelson Biographical." https://www.nobelprize.org/prizes/physics/1975/mottelson/biographical.

Chapter 14: James Watson

Bryson, Bill. *A Short History of Nearly Everything.* New York: Broadway Books, 2003.

"Decoding Watson." *American Masters.* Directed by Mark Mannucci. 2019. PBS.

"Discover Dialogue: Geneticist James Watson." *Discover Magazine,* July 2003.

Simmons, John. *The Scientific 100.* Secaucus, NJ: Citadel Press, 1996.

Watson, James D. *The Double Helix.* New York: Atheneum Publishers, 1968.

———. *Genes, Girls, and Gamow: After the Double Helix.* New York: Vintage, 2003.

Watson, J.D., and F.H. Crick. "A Structure for Deoxyribose Nucleic Acids." *Nature* 171 (1953): 737–38.

Chapter 15: Ei-ichi Negishi

Nobel Prize. "Ei-ichi Negishi Biographical." https://www.nobelprize.org/prizes/chemistry/2010/negishi/biographical.

Chapter 16: Ferid Murad

DePauw University Video Archives "2004 – Nobel Prize Winner Ferid Murad '58 Receives Honorary Doctorate from DePauw U." January 14, 2013. YouTube video, 9:39. https://youtu.be/WlfiiY1qRx4.

"Ferid Murad '58." *DePauw Magazine*, Spring 2019, 14–15.

The George Washington University. "GW Welcomes to Faculty Nobel Laureate, Dr. Ferid Murad." January 12, 2011. YouTube video, 27:04. https://youtu.be/7L2Yg2veNCO.

Nobel Prize. "Ferid Murad Biographical." https://www.nobelprize.org/prizes/medicine/1998/murad/biographical.

Chapter 17: Richard Schrock

Nobel Prize. "Richard R. Schrock Biographical." https://www.nobelprize.org/prizes/chemistry/2005/schrock/biographical.

Chapter 18: Eric Wieschaus

Hutchison, Lynne. "Trudi Schüpbach and Eric Wieschaus: A Shared Passion for Nature's Truth." Lens: A New Way of Looking at Science, August 2008.

Nobel Prize. "Eric F. Wiescaus Biographical." https://www.nobelprize.org/prizes/medicine/1995/wieschaus/biographical.

Wieschaus, Eric F. "American Society for Cell Biology Member Profiles." Newsletter of the American Society for Cell Biology, July 15, 2007. www.ascb.org.

ABOUT THE AUTHOR

DUANE S. NICKELL is a retired physics teacher in Indianapolis. He is a past president of the Hoosier Association of Science Teachers and a winner of the Presidential Award for Excellence in Science and Mathematics Teaching, the nation's highest honor for science and mathematics teachers. His previous books include *Guidebook for the Scientific Traveler: Visiting Astronomy and Space Exploration Sites Across America* and *Guidebook for the Scientific Traveler: Visiting Physics and Chemistry Sites Across America.*